ROCKETEERS

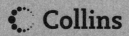
Smithsonian Books

Collins
An Imprint of HarperCollinsPublishers

ROCKETEERS

How a Visionary
Band of Business
Leaders, Engineers,
and Pilots Is Boldly
Privatizing Space

MICHAEL BELFIORE

ROCKETEERS. Copyright © 2007 by Michael Belfiore. All rights reserved. Printed in the United States of America. No part of this book may be used or reproduced in any manner whatsoever without written permission except in the case of brief quotations embodied in critical articles and reviews. For information address HarperCollins Publishers Inc., 10 East 53rd Street, New York, NY 10022.

HarperCollins books may be purchased for educational, business, or sales promotional use. For information please write: Special Markets Department, HarperCollins Publishers Inc., 10 East 53rd Street, New York, NY 10022.

FIRST EDITION

Designed by Nicola Ferguson

The Library of Congress Cataloging In Publication Data
　Belfiore, Michael P., 1969–
　　Rocketeers : how a visionary band of business leaders, engineers, and pilots is boldly privatizing space / Michael Belfiore. — 1st ed.
　　　p. cm.
　　Inlcudes bibliographical references and index.
　　ISBN: 978-0-06-114902-3
　　1. Rocketry—United States. 2. SpaceShipOne (Spacecraft) 3. Aeronautics—United States—Records. 4. Rutan, Burt. 5. Space travelers. 6. Outer space—Civilian use. I. Title.

TL781.8.U5B45 2007
338.0919—dc22

2007014987

07　08　09　10　11　ID/RRD　10　9　8　7　6　5　4　3　2　1

FOR AMELIE, MY X PRIZE BABY

Contents

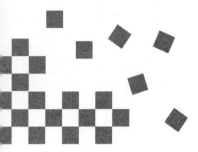

ACKNOWLEDGMENTS

This book began as a bare outline in 2003. From there to the book you hold in your hands was a long journey, during which I received a lot of help and encouragement. I offer my sincere thanks and gratitude to my wife, Wendy Kagan, for supporting me in this particular mad scheme from the beginning; to Jeff Foust for his advice and continuing good example; to Brian Feeney for giving me that first in-person interview; to the magazine and news editors who gave me assignments to cover many of the subjects of this book, especially Eric Adams, Mark Jannot, and all the good people at *Popular Science*; to Jeff Davis for introducing me to my extraordinary agent, Linda Loewenthal; to my editor at Smithsonian Books, Elisabeth Dyssegaard, who just plain gets it; and to all the rocketeers who have opened their doors and patiently explained rocket science to me: thank you!

ROCKETEERS

PROLOGUE

Full Circle

Mojave Airport, Mojave, California, October 4, 2004.
It was not yet noon, but the desert sun blasted down
relentlessly, painfully bright to one used to the less
sunny skies of upstate New York. I'd been on the
go since 3:30 that morning, when I'd been roused
from my hotel room by Gina Keating, the Reuters
staffer I'd been paired with to cover the winning of
the X PRIZE—$10 million for the world's first private
spaceship. I was sun dazzled, sleep deprived, and
hopped up on adrenaline. Twenty thousand people
pressed against the fences along the flight line at a
public viewing area. In the media section, dozens of
TV trucks pointed satellite dishes at the sky while
reporters spun the space news story of the year, each
sheltered from the sun in an identical, open-sided tent.

Brian Binnie, test pilot for Scaled Composites, had taken off from the airport at dawn in a bullet-shaped rocket plane attached to the belly of a gangling jet whose long, spindly tail booms and landing struts made it look like a gigantic insect. Now he was circling to launch altitude at 47,000 feet. After two hours in flight, he was almost there.

Keating and I had filed our first story and were heading back through the media area to our spots along the fence to watch Binnie light the rocket when I bumped into Eric Adams, aviation editor at *Popular Science*. Before heading west I'd just about talked him into assigning me a feature on the first private space station, now under development by hotelier Robert Bigelow outside Las Vegas. Jim Benson, CEO of SpaceDev, approached us, looking a bit furtive, with a SpaceDev ball cap pulled low against the sun. His company was a contractor working on the rocket engine of the spaceship winging ever closer to launch above us. He'd found a gap in the fence separating the media from the VIP viewing areas, and he'd been moving back and forth. Launch time was drawing near; he wanted to get back, and we could come along if we wanted to.

The VIPs had something the press didn't—a giant TV screen playing video from cameras mounted on the spaceship. While we waited for launch, Eric schmoozed with Steve Fossett, who was soon to make history as the first pilot to fly solo around the world nonstop. I said hello to Brian Feeney, resplendent in a bright yellow flight suit festooned with sponsor logos. He'd lately been making news by claiming to be close to a launch of his own, in a homemade spaceship dangling from the bottom of the world's largest reusable helium balloon. Feeney was a self-taught industrial designer from Toronto, and a perfect example of the can-do spirit of an emerging breed of space entrepreneurs. As it turned out, he was nowhere close to reaching space, but his bravado made good copy for the papers.

Binnie's spaceship dropped from its carrier plane, and a cheer went up from the crowd. Binnie lit the rocket engine, to more cheers. We saw it on the TV screen as a dirty orange flame blasting out the back of the spaceship along with dense black smoke. With our naked eyes we saw it as a single bright contrail scratching the pale blue sky over the Tahachipi Mountains to the east like a swiftly moving chalk line, heading straight up at astonishing speed. It rose between the mountain peaks and the still-rising sun like a force of nature, awesome to behold. Our spirits rose with Binnie on ninety seconds of flame, and when the engine cut off and the contrail began to drift slowly in the winds of the high atmosphere, we went weightless with him, craning our necks back and imagining what it must feel like to cast loose the bonds of gravity, to see the curvature of the Earth against an absolutely black daytime sky, and to feel the deepest, most abiding sense of peace and silence known to man or woman. It seemed I'd spent my whole life preparing for this timeless moment.

I was six years old when I first read about rockets sending people into space. In the public library of Canoga Park, California, I picked up a young adult novel called *Rocket Ship Galileo* by Robert Heinlein. Published in 1947, it tells the story of a trio of high school kids and a nuclear physicist who build a spaceship in the New Mexico desert and fly it to the moon. One of its full-page drawings showed people reclining on acceleration couches in front of instrument panels festooned with dials and buttons and levers. And out the big windows above the instruments: a black sky filled with stars. From that moment on I was fascinated by the idea of space travel.

Little did I know that by the year I picked up Heinlein's book, 1975, the glory days of the first space age were over and done. Only twelve men had walked on the moon. The last of them, the crew of *Apollo 17*, had packed up and left three years before. The manned space program of the U.S. National Aeronautics and Space Administration had already begun its long, slow slide into a bureaucratic morass from which it might never recover. I didn't

3

even know that people had already walked on the moon. All I thought was "Imagine: *people on the moon!*"

Two years later, in the library at my local Boys' Club in Santa Monica, I happened to glance up at the wall above a set of bookshelves I must have passed a hundred times. There hung a picture of a man in a bulky white pressure suit standing on a gently undulating plain against a black sky. I recognized it immediately from my reading of *Rocket Ship Galileo* as a man on the moon. But this was no drawing. In the days before computer-generated imagery, there was no other explanation for such a photo-real image: people had actually been to the moon! But I had to make sure. I snagged a passing counselor. "Excuse me, is that . . . ?"

"A man on the moon, yes," he said impatiently.

Eight years after the fact, the first moon landing hit me with the force of revelation, as it had countless other people when it happened. Time stood still, every detail of that moment etching itself into my memory. Light blazed in from the windows to my left, where an aquarium containing a fish-eating sea anemone, a source of endless fascination to us boys, burbled quietly. The nostril-tickling scent of fresh-cut wood and sawdust drifted in through the open library door from the shops on the other side of the building, along with the clacking and thumping of foosball and air hockey tables closer at hand. A man on the moon. It had actually happened! I memorized every detail of the picture on the wall—Buzz Aldrin's spacesuit with its gold visor and Neil Armstrong and a landing strut of their spaceship reflected in it, the air hose attached to his chest snaking back under his arm to his backpack, the black moondust on his boots, the black-as-space shadows he cast in the low sunlight of lunar morning.

In the years that followed, I read all I could about NASA's moon missions and all the Mercury and Gemini missions that had led up to them. Inspired by Heinlein and Aldrin, I decided to become a writer of space stories. Not much new had been done in manned space travel since the Apollo days, but imagination was the only limit to writing science fiction instead of science fact. I was writing science fiction stories and sending them to magazines by the time I was fifteen, and I sold my first one coming out of college in 1991. For the next thirteen years I pursued dual careers in acting and science fiction writing. I found work as a technical writer, interviewing engineers to put the complex machinations of industrial software into clear, simple language for user manuals, but science fiction never did pay my rent. Trouble was, not so many people read stories about space travel any more, and those who did could keep in print only a few hundred writers.

NASA had struck the first blow in the late 1950s, killing off a whole class of science fiction stories by convincing the general public that only big government programs spending billions of dollars could send anyone into space. Before that, stories like *Rocket Ship Galileo* were legion; amateurs building their own rides to space were seen as just a near-future extension of the way countless other industries had gotten started. Who, after all, had built the first automobiles and airplanes but talented and determined amateurs? A lot of science fiction's charge came from the idea that it could actually happen, and soon—maybe even to the people reading it. But by the 1960s, those kinds of stories seemed hopelessly quaint and outdated. Amateurs and small companies, the writers thought and readers agreed, could no more build spaceships than interstate highways. No matter, though; NASA was heading to the moon, cost be damned, because the money was there for it—money that many saw as a well-spent down payment on humanity's future in space. Science fiction stayed popular in the 1960s by reflecting the times with tales of governments and megacorporations seeding the cosmos. Readers could still imagine themselves flying the spaceships of these stories, settling other planets, and generally participating in a space-faring civilization. This was heady stuff, full of hope and optimism for the future, punctuated by the real-life adventures of NASA's astronauts flying ever higher and ever faster in the great moon race against the Soviet Union.

Space fever reached its pitch at the end of 1968, with the fall release of Arthur C. Clarke and Stanley Kubrick's film *2001: A Space Odyssey* and the Christmas Eve flight of *Apollo 8,* the first spaceship to send people to the moon. *2001* captured 1968's optimism about humanity's future in space perfectly, showing our push to the stars as a direct extension of our evolution as a species. It depicted near-future spaceships, an orbiting space station, and a moon base so rigorously conceived and yet so perfectly ordinary, complete with corporate logos and receptionists, that they just *had* to come true. And it gave audiences a taste of the unimaginable riches that might await explorers of deep space, far beyond the Earth-moon system, where advanced beings might only await our collective maturity before welcoming us into a galactic community.

Although *Apollo 8* didn't land on the moon, it took humans on their first voyage to another world and thus had perhaps an even greater impact on humanity's collective psyche than the *Apollo 11* landing seven months later. The *Apollo 8* crew's photographs of a crescent Earth rising above a moonscape and their accompanying live TV broadcast beamed back to Earth a message of hope from outer space that was badly needed after a year packed with bad news—race riots, the assassinations of Martin Luther King, Jr., and Robert F. Kennedy, and the horrors of the Vietnam war brought home with the Tet offensive and the My Lai massacre. Pulitzer Prize–winning poet Archibald MacLeish summed up that message of hope on Christmas day in the *New York Times:* "To see the earth as it truly is, small and blue and beautiful in that eternal silence where it floats, is to see ourselves as riders on the earth together, brothers on that bright loveliness in the eternal cold—brothers who know now they are truly brothers." Recognizing *Apollo 8* as the most inspiring news story of the year, *Time* magazine named the crew as its "men of the year." Shooting the moon had always been a metaphor for attempting the impossible. Yet, we human beings had done it. What other "impossible" goals might we achieve? Could we wage peace? Cure disease? Feed the poor? Create entirely new forms of music, art, theater? Influential movements sprouted up during the 1960s to advance these grand ambitions and more. It's no coincidence that all this Renaissance-like activity peaked with the moon race; as our physical horizons expanded, so did our minds in all sorts of wonderful and unexpected ways.

Unfortunately, Neil Armstrong's first steps on the moon on July 20, 1969, marked the beginning of the end of the first space age. With the winning of the moon race, the U.S. national space program lost its direction. Congress cut back its funding and left it without the all-consuming purpose that had fueled it through the 1960s. It turns out, that's *all* NASA had set out to do—shoot the moon, by any means necessary. It got there with single-use rockets and spaceships, and afterwards dismantled the infrastructure it would have needed to build more. The space agency coasted through the 1970s, turning the last of the moon rockets into *Skylab*, the first U.S. space station, and sending three manned missions to it with the last of the Apollo spaceships. The space shuttle, which had been conceived to service space stations, didn't materialize until the 1980s, over budget and very late to the party. Without spaceships to provide the regular boosts it needed to maintain its orbit, *Skylab* crashed to Earth two years before the shuttle's debut, making the shuttle a ship without a port before it even launched.

Meanwhile, public interest in space travel faded along with the idea that we were living in a space age. Nothing really new was happening; no new frontiers were being explored, at least not by people. Astronauts flying in circles around the Earth servicing satellites and fulfilling obscure scientific missions didn't inspire people the way visiting new worlds did. And since the big government space monopolies couldn't effectively build on the legacy of the moon race, most people assumed that no one could. It seemed that humanity really was destined to be Earthbound after all, traveling no higher than the shuttle's low Earth orbit.

The literature of science fiction retreated to tales of post-apocalyptic future Earths, and when the personal computer revolution hit in the 1980s, it took refuge in inner space, exploring virtual worlds through characters only too eager to escape from reality. As with tales of amateurs building rockets to the moon, optimistic stories about a united humanity finding itself in the stars seemed naive and outdated. The popularity of science fiction stories and novels contracted as readers fled to lighter escapist fare like sword-and-sorcery novels.

By 2001, the year for which Clarke imagined hotel-style space stations and a permanent moon base, only three major science fiction magazines clung to a precarious existence among the poetry and literary magazines on the bottom shelves of only the biggest book stores' magazine racks. Still, I soldiered on with my stories.

That is, until I found out about the X PRIZE. Competing for the prize were real people, engineers at small companies as well as amateurs, building honest-to-God spaceships, just as in *Rocket Ship Galileo*. Turns out a lot of other people who had been inspired by the science fiction of old had gotten tired of waiting for government programs to give them their rides to space. Spurred on by X PRIZE founder Peter Diamandis and his $10 million cash prize for the first private craft to send a pilot and two passengers into space twice in two weeks, these folks were building their *own* rides. This was better than science fiction; this was for real, with real people to talk to, and real machines to climb inside.

On December 17, 2003, the 100th anniversary of the Wright Brothers' first powered flight, a spaceship built by an X PRIZE competitor with the unassuming name of Scaled Composites flew under rocket power for the first time. The ship went supersonic, something no privately built craft had done before, and pilot Brian Binnie rode to a peak altitude of 67,800 feet before returning to Mojave safely. *Whoa!* I thought; *These guys are really going to do it!* I knew that soon Scaled would fly a manned test flight to space in preparation for the X PRIZE attempts. That flight, even more than the X PRIZE flights, would change everything because it would be the one to shatter NASA's monopoly on spaceflight. It would be big news, I knew, and if I could just get to Mojave with an assignment from a news outlet—any news outlet—I could ride *SpaceShipOne*'s twin tail booms to a new chapter in history.

The *New York Post* gave me the break I needed, and there I was on the ground at Mojave on June 21, 2004, as Scaled pilot Mike Melvill became the world's first commercial astronaut. I'd found my calling—in the bright sunshine among my fellow writers and people actually doing things in the world. Best of all, I was writing stories that would have been right at home in old issues of *Amazing Stories* or *Astounding Science Fiction.*

Keating and I attended the press conference following Binnie's X PRIZE–winning flight, and then we headed back to the filing center to send Reuters our final report.

My cell phone warbled. "This is Robert Bigelow's office calling. Can you take a call from Mr. Bigelow?"

Eric Adams had agreed to give me a *Popular Science* feature on Bigelow's commercial space station project, but on one condition: I had to spend no less than three days with Bigelow and his engineers at their North Las Vegas research and development center. I'd already spent a couple of hours there on assignment with Reuters, so that didn't seem like such a big deal to me. It was a big deal to Bigelow, though. As I later found out, he'd never given a reporter that kind of access to his program or even allowed himself to be photographed for print. "Why do you need so much time?" he wanted to know.

"Good question. Let me call my editor and ask him. Call you right back."

In the midst of the negotiations, Bigelow hit me with some news. He had chosen this day, the day the X PRIZE was won, to announce the next great space prize. He was going to offer $50

9

million for the first private spaceship to send people not just out of the atmosphere and straight back down, like *SpaceShipOne*, but into orbit: a distinctly more difficult proposition.

SpaceShipOne was the only American craft to fly astronauts into space in 2004 (NASA's shuttles remained grounded following the destruction of *Columbia* on reentry the previous year). At last, space travel was no longer just the domain of prohibitively expensive government programs subject to political whim. Now it was just like any other business that could be developed into a thriving industry.

Scaled Composites was but one of many such small companies building spaceships on a relative shoestring. A company with only 135 employees, Scaled developed *SpaceShipOne* for a mere $25 million, the price of a low-budget Hollywood film. Robert Bigelow plans to spend $500 million on the first commercial space station. That's no more than NASA spends on a single space shuttle mission. Space Exploration Technologies (SpaceX) CEO Elon Musk took the first step in his own manned space program by building a satellite launcher for just $100 million.

Even NASA had to admit that small entrepreneurial companies might just provide a viable alternative to pumping billions of dollars into big aerospace companies for hardware that often never materializes. Pushed by the grounding of the shuttle fleet into the humiliation of having to buy rides from its former archrival, the Russian space agency, NASA put out a request for proposals in 2006 for privately owned spaceships that could service the International Space Station. It committed $500 billion to help entrepreneurial companies develop those ships, and then to buy rides on them when they were completed.

By then, Scaled Composites was hard at work on a fleet of larger versions of *SpaceShipOne* that had been ordered by Virgin Atlantic Airways owner Richard Branson for a daring venture called Virgin Galactic. Virgin sold the first hundred tickets for suborbital joyrides on the new spaceships even before they were completed, and it planned to send more people into space during its first year of operation than had flown during all the previous years of manned spaceflight combined.

We're at a point in history analogous to the beginning of the personal computer boom, when visionaries like Steve Jobs and Bill Gates were building their first computers and writing their first programs in their garages. Or the beginning of the twentieth century, when Henry Ford opened his first assembly line. Tourism is the first market for the new spaceflight industry, as thousands of people with the dream to see the Earth from space for themselves sign up for rides on suborbital spaceships, which will become increasingly affordable. Eventually those ships will travel not just straight up and down on sightseeing jaunts, but also across continents and oceans much faster than the supersonic Concorde passenger airplanes, and with far less of the noise that kept Concorde from operating on overland routes. That development alone could alter the travel industry as thoroughly as did the jet plane. Tourists will also provide some of the first markets for orbital ships that will rocket them to vacations on commercial space stations. Commercial flights beyond the Earth's orbit won't be far behind; in 2005 the increasingly entrepreneurial Russian space program cut a deal with Florida-based Space Adventures to send two tourists and a professional cosmonaut on a flyby mission around the moon. The major hardware for the mission already exists; all that's needed now is to find two people willing and able to pony up $100 million each to pay for it.

This, then, is the story of the second space age, starting with the first prize for manned spaceflight and the white-knuckle test flights of the first commercial spaceship, continuing through the advent of the first commercial orbital ships and space stations, and taking a good look at what the future in space might bring. This is fundamentally a hopeful story, about people of all ages and political persuasions daring to look *up* for answers to our

Earthly problems, about people who believe that the human species has yet to realize its potential, that our bonds are far stronger than our differences, and that, as MacLeish said, we all of us
are members of the same family "on that bright loveliness in the
eternal cold."

1

SPACE OR BUST

Peter Diamandis
and the X PRIZE

Princeton, New Jersey, spring 1994. It was a perfect day for flying. Clear and smooth blue skies, with a light breeze from the west. Even better, it was Sunday, and space entrepreneurs Peter Diamandis and Gregg Maryniak had a rare day off together. Diamandis and Maryniak had just come off a long hard run at starting a satellite-launching business together, an effort that had proved unsuccessful. Actually, Maryniak was to observe less charitably years later, "we lost a shirt each."

For Maryniak, a pilot from the age of sixteen, there was no better way to get some much-needed perspective than by renting a Cessna at Princeton Airport and taking to the air. Diamandis, who'd put in twenty-five to thirty hours of flight time toward his own pilot's license, couldn't agree more, though he rarely seemed able to find the time to fly simply for enjoyment.

Their plane for this day was a Cessna 172, like the others at the airport, a gracefully aged quarter century old. This was flying as it was meant to be, thought Maryniak as he climbed into the pilot's seat: controls little changed from aviation's golden age, simple mechanical indicators, and a push/pull throttle jutting from the instrument panel. Flight, distilled to its purest elements in a slow, low-flying plane that lets you feel the wind pushing back against the airframe, gives you views on all sides, and even, with a gentle roll to the left or the right, allows you to look straight down on treetops and highways.

Diamandis got in the copilot's seat, and Maryniak's ten-year-old daughter climbed in the back. They took off to the west and then turned east toward Raritan Bay. The steely gray waters of the Atlantic Ocean and the New Jersey shore rolled into view below them, and in the distance to their left, the glass towers of Manhattan's financial district flashed in the sun. Maryniak steered the plane north, keeping Manhattan on their right, overflying New York Harbor and the mouth of the Hudson River.

To their left the cliffs of the New Jersey Palisades gave way to the green, sun-dappled mountains made famous by the Hudson River school of painters more than a century before. It didn't take an extreme stretch of the imagination—a squint to one direction rather than another at the mountains rising on either side of the river—to see the river and its valley as Henry Hudson had found it nearly four hundred years before, unspoiled by industry, densely forested with hardwoods hundreds of years old, the domain of vast flocks of blackbirds and wild turkeys and light-footed Native Americans hunting deer, a place of mystery and adventure and vast, untapped natural resources: the jewel of the New World.

Soaring above it all, Diamandis got excited about finally finishing his flight training. "Maybe this time I'm really going to do it," he told Maryniak.

"Isn't this just magnificent?" Maryniak agreed.

That day was still fresh in his mind a week later when Maryniak passed a bookshop with his wife and daughters near their home in Princeton. There he found a copy of one of his favorite books, *The Spirit of St. Louis,* by Charles Lindbergh, and decided to give it to Diamandis as a present. The book was a chronicle of Lindbergh's epoch-making solo flight across the Atlantic Ocean in 1927. But more than that, its lyrical recounting of the golden years of aviation was a siren call to all who dreamed of flight. Maryniak thought that maybe the book, along with the memory of the perfect day of flying they had shared, would inspire Diamandis to finish his flight training.

The book inspired Diamandis, all right, but not in the way Maryniak expected. Perpetually on the move, always scheduled to the hilt with his various business ventures, Diamandis let months go by before he found the time to read *The Spirit of St. Louis*—when he settled into a winter vacation at his parents' house in Florida. Once he did start reading, though, he couldn't stop. "I just spent the entire vacation in my room reading this book like my life depended on it," he later recalled. As Maryniak had thought, Diamandis was inspired by Lindbergh's recounting of aviation's barnstorming years after World War I. But what really blew him away was the book's revelation of Lindbergh's motivation for flying solo across the Atlantic. Like most people with a passing familiarity with Lindbergh's feat, Diamandis had assumed that Lindbergh had made the journey in May 1927 simply as an enormous personal challenge. In fact, he did it to win a prize.

The prize was $25,000 in cash, offered in 1919 by a French-American hotelier named Raymond Orteig "as a stimulus to courageous aviators . . . to be awarded to the first aviator of any Allied country crossing the Atlantic in one flight, from Paris to New York or New York to Paris. . . ." When Lindbergh took off from Roosevelt Field on New York's Long Island on May 19, 1927, two other Orteig competitors were on the ground preparing for transatlantic flights of their own. One of them was the favorite to win the prize: Commander Richard E. Byrd, the well-financed aviator who had made headlines for overflying the North Pole earlier that month. Lindbergh's success was due in part to his daring to take off into bad weather in a calculated risk based on years of experience flying the mail on cross-country flights. Significantly to Diamandis, none of the three Orteig competitors who eventually made the transatlantic crossing used new or exotic technology to do so; all flew airplane and engine designs already in use at the time. The feat was made possible by acts of courage rather than by exotic new technologies.

A close competition featuring an underdog pitted against the star aviator of the day, a cash prize, and feats of derring-do—all contributed to the American public, in the words of Lindbergh biographer A. Scott Berg, "behaving as though Lindbergh had walked on water, not flown over it." Lindbergh became the most photographed man in the world, the first megacelebrity of the twentieth century. Thirty million Americans showed up to see Lindbergh on a cross-country speaking tour following his transatlantic flight—fully a quarter of the entire population of the United States—and uncounted schools, mountains, lakes, and streets were renamed to honor him.

The result of all this adulation was to catapult the aviation industry to the center of public consciousness as the next great industry. The year 1927 marked the breakout year of commercial aviation in the United States, the beginning of what came to be called the Lindbergh boom. In April, the month before Lindbergh's flight, 97,000 pounds of mail flew on airplanes. In September, that figure was up 50 percent, to 146,000 pounds. The number of applicants for pilots' licenses tripled that year, and the number of licensed airplanes quadrupled. A writer for *Forbes* magazine observed in 1927 that "Lindbergh's significance to business seems greater than that of any mercantile or financial magnate on either side of the Atlantic." Perhaps most tellingly, the number of passengers flying on airlines in the United States increased from 5,782 in 1926 to an astonishing 173,405 by 1929.

The crucial piece of information about Lindbergh's motivation for flying the Atlantic brought a lightning flash of inspiration to Diamandis. He saw that such a prize, updated and offered in the present as a *space* prize, might be just what was needed to bring space travel to the general public, to jump-start a commercial space industry. Diamandis put down the book. He took a deep breath. And then he started reading again from page one, this time taking extensive notes in the margins. By the time he closed the book, the idea of a suborbital space barnstorming prize had fully formed in his mind.

Diamandis wasn't wealthy, so he couldn't personally award the 1990s equivalent of the $25,000 that Orteig had offered from his personal fortune. He didn't let that discourage him, though—he would find a financier and convince him or her to put up the prize money. He had no doubt of his ability to do that as long as the amount of money he settled on was within reason. But that still left the problem of what to call his prize. *I don't have an Orteig yet,* he thought, *so I'm going to call it X as a variable for the name of the person.* The letter X was also the designation of a long line of experimental aircraft built by the United States government, including the X–15, America's first manned spacecraft, so it was fitting for that reason as well. Next, Diamandis had to consider just what would be a large enough incentive to offer as a space prize. He knew it had to be in the millions of dollars for someone with any chance of winning the prize to take it seriously, yet low enough for Diamandis to have a hope of actually raising it. What was the magic number? Five million? Ten? Twenty? And then it came to him: *Ten is the Roman numeral X! That's perfect!* "It was one of those things," he recalled later, where "you reach across time and space and touch a powerful idea that you know is going to work."

Peter Diamandis was born in the New York City borough of the Bronx on May 20, 1961, just a couple of weeks after Alan B.

Shepard became America's first Project Mercury astronaut. Diamandis was an impressionable eight years old when Armstrong and Aldrin set foot on the moon, and he thrilled to the exploits of all the moon walkers through 1972. He was ten years old when he fully appreciated the significance of the moon landings. It came in a flash-bang of understanding while he was sitting in his fifth-grade class listening to a lesson on the planets. "I didn't understand why," he recalled later, "but I knew in that moment that my 'mission' in life would be the exploration and development of space. From that day forward, everything I did or dreamt about had a flavor of space." The wildly popular television show *Star Trek*, with its vision of a future in which members of all nations worked side by side in a United Federation of Planets, further inspired Diamandis with the potential for space travel to advance the causes of peace and prosperity for all.

None of which seemed all that far-fetched, considering the leap that his father had made years before. Harry Diamandis was born the son of an olive oil merchant in a tiny village on the Greek island of Lesbos. Harry's dream was to become a doctor, so at the age of eighteen he became the first person in his family to leave the island—to go to Athens to take his medical board exams. Eventually his dream took him across the world to New York City, about as far removed from his birthplace as he could get, there to set up a successful practice as an obstetrician. That example taught Diamandis that he could accomplish anything he set his mind to.

Initially, Diamandis set his sights on becoming an astronaut with NASA. A natural enough choice, considering that only NASA and its Russian counterpart, the Russian Federal Space Agency, had ever sent anyone into space. It was also a natural for Diamandis to study medicine, since his parents expected him to become a doctor. Then, too, medical researchers made up the bulk of NASA astronauts who weren't test pilots, so he could fulfill both sets of expectations with one degree. Diamandis attended college at the prestigious Massachusetts Institute of Technology (MIT), there to study molecular biology and fulfill his undergraduate requirements in preparation for medical school. For good measure, he stayed at MIT for a master's degree in aerospace engineering, another credential that would look good to NASA, and went on to Harvard University for medical school.

While in graduate school in the late 1980s, Diamandis met actual astronauts for the first time. And he was in for a shock. Astronaut Byron Lichtenberg told Diamandis that the chances any given applicant would actually get into the NASA astronaut corps were a miserable one in a thousand. Well, Diamandis never expected it to be easy to realize his dream. But Lichtenberg further explained that only about half of all NASA "astronauts" actually got to fly. The rest flew desks in various administrative jobs, hoping against hope for the chance to rocket into space. And even those lucky one in two who got a flight went up, on average, only a couple of times in a decade—and that might well be all the spaceflight experience they got before they had to retire. "That's not my vision of spaceflight," said Diamandis. "I want to go when I want to go, just like if you're a scuba diver or a mountain climber." Finally Lichtenberg explained to him that NASA astronauts had to be expert players of office politics and follow every NASA rule and regulation to the letter; otherwise they had no chance at all of achieving escape velocity from the fierce gravitational pull of their desks. That didn't sit well with Diamandis at all. "I realized I was more of a rebel than that," he recalled later. Then Diamandis made the kind of imaginative leap that had allowed his father to go from a remote Greek island village to treating OB-GYN patients in New York City. "Well," he said, "it looks like I'm going to have to get into space in a different way than just flying for NASA. I'm going to have to do this thing privately." He didn't work out just how he was going to pull that one off until he read *The Spirit of St. Louis* seven years later. Once he figured it out, though, he didn't waste another minute. He started making phone calls.

. . .

Like many of the people Diamandis called to sound out about the X PRIZE idea, Gregg Maryniak was skeptical. But foremost on Diamandis' mind was an axiom first put to words by Arthur C. Clarke in 1968: "All revolutionary ideas . . . pass through three stages, which may be summed up by these reactions: (1) 'It's crazy—don't waste my time.' (2) 'It's possible, but it's not worth doing.' (3) 'I always said it was a good idea.'" Diamandis expected to encounter resistance, and he was prepared to overcome it.

Maryniak didn't dismiss the idea out of hand; he had actually been thinking along the same lines himself. The previous year, in fact, he had cited Lindbergh in a paper about the challenge of making space travel truly commonplace as "an example of a person who dramatically changed the way people thought about aviation with a small-scale effort," with the implication that an equally influential person could do the same for spaceflight. He, like Diamandis, believed that the main barrier to the large-scale commercialization of space was not that the technologies involved were too difficult or expensive to implement, but rather the *perception* on the part of investors that this was so, along with the corollary belief that space travel was best left to big government programs. Maryniak also knew from his reading of history that aviation prizes, of which the Orteig prize was only the most influential, had done much to advance the art and science of aviation before World War II. He had a different concern about the X PRIZE. "Peter," he told Diamandis, "you know and I know that ten million bucks is chicken feed." Maryniak didn't think the X PRIZE could attract serious competitors for what amounted to pocket change as far as spaceflight was concerned. Diamandis argued that it could with the right set of rules. The two friends spent the better part of a year working out just what those rules should be, and by the end of that time, Maryniak was a convert—the first of many—to Diamandis' idea.

To work out the rules for the X PRIZE, Diamandis and Maryniak, now joined by Byron Lichtenberg, considered the history of aviation prizes. Unlike the beginning of spaceflight, early aviation was dominated by tinkerers, small businesses, and garage inventors rather than governments, and many of the advances they made were fueled by cash prizes offered by some of the premier institutions and business leaders of their day. The London *Daily Mail*, the *New York Times*, and William Randolph Hearst offered prizes for accomplishing everything from a mere fifteen minutes of sustained flight in 1908 (10,000 French francs offered by Jules Armengaud and won by Henry Farman) to the first nonstop transatlantic flight, offered not by Orteig but by *Daily Mail* publisher Lord Alfred Northcliffe in 1913. The duo of John Alcock and Arthur Whitten Brown won that prize (£10,000) in 1919 by flying a World War I bomber the shortest distance possible across the Atlantic, from Newfoundland to the Irish coast.

Those early aviation prizes offered incentives for small incremental steps toward faster and farther-flying aircraft. Alcock and Brown could never have crossed the Atlantic without such intermediary steps as the first crossing of the English Channel (by Louis Blériot in 1909 to win £1,000 offered by the *Daily Mail*) and the first flight from New York City to Albany, the state capital of New York, 134 miles away by air (by Glenn Curtiss in 1910 to win $10,000 offered by the New York *World*). To compete for these prizes, engineers and aviators built and flew an ever-expanding variety of airplanes and engines, perfecting the state of the art through competition, which eventually led to the development of planes that were safe enough and practical enough for ordinary people to fly simply to get from one place to another.

What Diamandis and Maryniak knew they needed, then, was to foster a baby step, as in the aviation prizes of old. Private efforts to build manned spacecraft in the past had all failed, they reasoned, because they attempted to do the space equivalent of flying the Atlantic: launching a craft into orbit without the necessary intermediary steps. To reach orbit, the space shuttle travels twenty-five times the speed of sound and climbs more than two hundred miles above the Earth's surface—a formidable task for any privately funded effort, and certain to cost many times the X PRIZE purse.

It seems obvious in retrospect, but no one had framed the problem in just this way before. It was twofold: First, what's a reasonable baby step for space travel that might actually have a good chance of succeeding? Second, what's the market for such a step, one that would sustain the development of the faster and higher-flying spaceships of the future? Diamandis and Maryniak hit upon the idea of suborbital spaceflight, a realm that had briefly been visited in the early years of the U.S. space program in the X–15 rocket plane and the Mercury space capsule, and then abandoned. It seemed a perfect place to aim the first space prize because it set the bar for winning the prize as low as possible while still requiring that the winner to reach space.

Space has no clear beginning; as any mountain climber knows, the atmosphere gets thinner and thinner the higher one climbs until it is too tenuous to support life, although it is certainly still present. Take off from the mountaintop in a high-altitude balloon, and the daytime sky darkens from blue to black even where there is enough air to support the balloon. The International Space Station, orbiting at an altitude of 220 miles, still encounters enough faint wisps of atmosphere to require a periodic boost from rocket motors to stay aloft. The U.S. Air Force puts the beginning of space at 50 miles straight up, a largely arbitrary altitude that makes a nice round figure. An internationally accepted figure puts the beginning of space at 100 kilometers (62 miles), another largely arbitrary round number. Diamandis and Maryniak wanted there to be no doubt about whether the X PRIZE winner reached space, so they chose 100 kilometers as their winning altitude. An X PRIZE–winning spacecraft would lack the velocity and the altitude to whiz around the Earth every ninety minutes the way the space shuttle did; it would simply blast up and out of the atmosphere for a quick look around, and then when it ran out of momentum at its peak altitude, or apogee, it would fall right back down again.

But even without reaching orbital speeds and altitudes, suborbital spacecraft would still allow their passengers to experience the high g forces of a powerful rocket boost, see the daytime sky turn as black as night, and watch the Earth curve beneath them in a view that stretched for hundreds of miles in every direction. On the horizon they would see the thin blue band of Earth's atmosphere—all that separated it from the infinite, airless void. They'd get a view that was most definitely out of this world, and for a few brief minutes they'd go weightless, becoming true astronauts. Diamandis and Maryniak were prepared to bet that enough people would pay enough money for that once-in-a-lifetime experience to make suborbital spaceflight a viable business proposition. Market studies conducted in 1993 and 1995 by Japan's National Aerospace Laboratory supported that idea, with 70 percent of Japanese respondents under the age of sixty reporting that they'd like to take a trip to space at least once in their lives, along with 60 percent of all Americans and Canadians surveyed. "We have met the payload of the future, and it's us!" said Maryniak. "It's people: carbon-based, self-replicating payloads that you can make using simple tools you have at home."

With the requirement that the winner launch a suborbital spacecraft having market potential, the rest of the X PRIZE rules fell into place. They were quite simple. The winner of the $10 million X PRIZE would have to meet these requirements:

1. Build a manned spacecraft, a *spaceship*, without any government funding. Government had had its shot at making space travel truly routine; now it was private industry's turn.

2. Launch three people in the spaceship to an altitude of 100 kilometers and return to Earth. This ensured that the winner would end up with a vehicle capable of flying two revenue-producing passengers in addition to a pilot.

3. Repeat step 2 with the same ship within two weeks. Diamandis and Maryniak wanted to foster the development of a spaceship that could be used not just once, but over and over again, on something approaching an airline-style schedule.

Although much easier than reaching orbit, suborbital spaceflight would still present a real challenge to the X PRIZE contestants. To reach 100 kilometers, the winning ship would have to travel Mach 3—three times the speed of sound. Even though that required only 1/25 the amount of energy needed to reach orbit, it was still faster than any private craft ever built. Although the craft could climb through lower altitudes by conventional means—jet engines or a balloon—it would need a rocket engine to make the final run to space. It would need a life support system to keep the pilot and passengers alive long enough to reenter the atmosphere, and some kind of maneuvering thrusters to orient it correctly for reentry. It would have to be sturdy enough to survive the heat and buffeting created by reentering the atmosphere at Mach 3, and it had to land in good enough shape to make the trip again within two weeks. The entire spacecraft and all of its engines and tanks had to be reusable. Only one craft in the history of spaceflight had ever met that last requirement: the U.S.-built X–15 rocket plane, and it had long ago become a museum piece. Still, Diamandis and Maryniak came to feel confident that the X PRIZE could be won without exotic new technology, and that $10 million, though it might not cover the winning vehicle's development costs, would at least provide substantial seed money for a winner, who would then be poised to start a business running tourist flights. Now all they had to do was come up with the prize money. That was Diamandis' department.

Peter Diamandis is one of those rare individuals blessed with both technical aptitude and a keen understanding of his fellow human beings. When I met him for the first time, in Mojave after *SpaceShipOne*'s maiden spaceflight, he greeted me with a heartfelt smile and a warm handshake. On subsequent meetings, however brief, he imparted the same warmth, soon greeting me by name. I saw him engaged in similar quick but familiar-seeming interactions with countless others at subsequent launches and space conferences. He gave everyone he met the feeling that he was genuinely glad to see them and that he wanted to hear what they had to say. It's a skill as uncommon as the ability to design a rocket engine that won't blow up on the pad, and perhaps just as important in starting a space business.

Diamandis is what Malcolm Gladwell described in his 2000 book *The Tipping Point* as a Connector. That is, he specializes in collecting people "the same way others collect stamps." And not just any people. "Connectors are important for more than simply the number of people they know," says Gladwell. "Their importance is also a function of the kinds of people they know." As an undergraduate, Diamandis started what has since become the premier international organization of students interested in space, Students for the Exploration and Development of Space. Later, as a graduate student at MIT, he launched an even more ambitious project called International Space University (ISU), now a fully accredited university with a master's degree program in space-related studies and a permanent campus in Strasbourg, France. Diamandis began both of those early ventures as ways to collect people interested in building a space-faring civilization. "Our vision was clear," he later said of himself and his fellow ISU founders. "We wanted a university where we could meet all of the future leaders of the space programs and forge a common vision of space. The ISU was our 'benign conspiracy.' We would bring together the best and brightest from around the world, create long-standing friendships, and come to a common shared vision of our future in space. Then, when the ISU graduates eventually enter positions of power in the decades ahead, we would be able to call on them to fulfill our common vision of space development." In this spirit, Diamandis set out to find donors for the X PRIZE.

At the urging of St. Louis Science Center director Doug King, Diamandis landed in the city that had given Lindbergh's bird its name. Like Lindbergh before him, Diamandis hoped to find a cadre of business leaders whom he could convince to fund his project as a way to promote St. Louis as a center of innovation. King introduced Diamandis to the man who was to become his main ally in this effort, public relations man Al Kerth. Kerth was a fellow Collector. And, as *St. Louis Post-Dispatch* reporter Eli Kintisch was later to put it, "In those days, if Kerth liked your idea, you were halfway there." Kerth sent out a letter to seventy-five of the wealthiest people in St. Louis informing them that "A new opportunity for St. Louis to make commercial aviation history has come to our attention." But it will "cost you a little bit of money."

"A little bit" turned out to be $25,000, Lindbergh's lucky number, the amount of the Orteig prize. Building contractor Ralph Korte was first to pony up. "It was an important thing to do for civic pride," Korte said later, "and my friend Al." On March 4, 1996, Diamandis and Kerth convened the first meeting of $25,000 donors at the same Racquet Club where Lindbergh had formed the original Spirit of St. Louis organization. Kerth called the group of high-rolling X PRIZE donors the New Spirit of St. Louis. By then its members included thirty of the city's most prominent citizens, including Enterprise-Rent-A-Car head Andy Taylor and his wife, Barbara. The two were lifelong aviation fans; Barbara's father had been a general in the Air Force, and Andy's dad had flown combat missions in World War II off the aircraft carrier USS *Enterprise*, which is where Andy got the name for his company. "The X PRIZE effort is good for mankind," said Andy. And, he added, "It's good for St. Louis."

Still, even with the cost of admission to the club of elite donors pegged at $25,000, Diamandis was a long way from $10 million. He needed all the help he could get in pushing those donations over the top, and perhaps even attracting that really big investor who would replace the "X" in "X PRIZE" with a proper name. Diamandis could think of no better way to do that than to get an actual descendant of Charles Lindbergh on board as a spokesperson.

Reeve Lindbergh, Charles Lindbergh's daughter, was less than enthusiastic about the idea when Diamandis approached her. But she knew someone who might be interested: her nephew Erik Lindbergh. Erik Lindbergh was a pilot, like his famous grandfather. That made him sound like a shoo-in to Diamandis. In fact, however, Diamandis had a tough sales job ahead of him. Erik agreed to meet Diamandis and X PRIZE cofounder Byron Lichtenberg for lunch in Seattle, where Erik lived, but that was as far as he would commit. It didn't help matters that Diamandis was chiefly interested in Lindbergh for the cachet his name carried.

Like most of the Lindbergh family, Erik tried as much as possible to live life on his own terms rather than in the shadow of his famous relative. Apart from being a general aviation pilot, the thirty-one-year-old was about as far removed from the world of speculative spaceflight ventures as one could get. For one thing, he was deeply concerned about the environmental impact of technology and felt strongly that Earth's problems ought to be well on the way to being solved before we move off planet. He also depended on the natural world for his living; he made rustic furniture from interesting bits of unprocessed wood. "Some people see things in clouds—I see things in wood," says his artist statement on his Web site. "As I contemplate my own tentative relationship with the earth, I find myself identifying with gnarled trees; our twisted trunks, knots and burls are a visible testament to the struggles we have lived through. . . ." The high-tech, science-fiction future that Diamandis wanted to bring about frankly left Lindbergh cold. "Can't we use ten million bucks better here on Earth?" he asked. "We have lots of problems we could fix with ten million bucks!" Yet, Diamandis had an opening. The X PRIZE project stirred old longings within Lindbergh that he hadn't considered since he was a boy, when he launched crickets in model rockets and thought about how exciting it would be to fly in space himself one day. And by the time Diamandis had worked his magic on him, Lindbergh was the latest member of Diamandis' growing legion of converts. "They got under my skin," Lindbergh later admitted of Diamandis and Lichtenberg. "They sort of found my hot buttons."

It wasn't a stretch at all for Diamandis to play the environment card. Concern for the environment was a fundamental part of his and Maryniak's inspiration for wanting to help humankind move into space. Both men were adherents of the ideas of Gerard O'Neill, a Princeton University physicist from the mid–1950s until his death in 1992. Those ideas, developed during the environmentally conscious 1970s, coalesced in 1976 with O'Neill's influential book *The High Frontier*.

The book laid out a blueprint for the exploration and development of space that would lead to building giant rotating (to simulate Earth-normal gravity) colonies that would free-float in space midway between Earth and the moon. These space islands would house tens of thousands of people in artificial worlds engineered with year-round sunshine, lush countrysides dotted with small villages, and even lakes and streams. Their inhabitants would travel through space on commuter spaceships to factories and farms floating nearby. The energy to run the colonies and the factories would come in an endless flood of raw power from that great nuclear dynamo that powers all life on Earth—the sun. Raw materials for the colonies and the factories would come from a source far richer in metals and other needed elements than Earth itself—the asteroids. O'Neill made a compelling case for the exploitation of the vast resources of the solar system as the only way to continue the unchecked growth of our civilization, now sharply limited by Earth's dwindling reserves. As O'Neill put it in his book, "Even if we were to excavate the entire land area of Earth to a depth of a half-mile, and to honeycomb the terrain to remove a tenth of all its total volume, we would obtain only 1 percent of the materials contained in just the three largest asteroids."

O'Neill's space islands would elevate the standard of living for all people on Earth as well as in the colonies. An endless supply of cheap solar power beamed down to Earth as microwave energy would solve the energy crisis we now face because of our dependence on fossil fuels, and it would allow even the billions of poor in the third world to enjoy standards of living approaching those of the first world. The industries that cause the most damage to the biosphere would be moved off planet, no longer forced to strip-mine Earth for ever scarcer resources and to pollute the atmosphere and ground water with dangerous by-products. The colonies would also help relieve the environmental pressures caused by the world population explosion in the most direct way possible, by providing virtually limitless room for expansion. Best of all, none of O'Neill's ideas required the application of yet-to-be-invented technologies. They could be realized with the technology of the 1970s.

O'Neill was one of Diamandis' heroes, a standard by which he measured his own efforts. (Maryniak was no less committed to O'Neill's vision of the future. In fact, from 1983 to 1992 he headed O'Neill's Princeton-based Space Studies Institute, a nonprofit organization charged with finding ways to turn O'Neill's space utopia into reality.) "Everything that we hold of value on this planet—metals, minerals, real estate, and energy—are in infinite quantities in space." Diamandis told anyone who would listen. "You know, the things that we fight wars over on Earth. The Earth is a crumb in a supermarket filled with resources. And our ability to access those resources, to give people on this planet increased standards of living, is critical." Erik Lindbergh was indeed listening during lunch that day in Seattle as Diamandis explained how the X PRIZE was the first step toward realizing a better future for planet Earth. "My eyes went wide open," Lindbergh later recalled. "I thought, 'This is very interesting, and it challenges my gut reaction, which is a general sort of environmentalist reaction to rockets racing around.'"

Lindbergh returned home with a creative fire ignited by the

possibilities of space. He looked at the bits of unfinished wood in his studio, and instead of rustic furniture, he suddenly saw "Martian smoke trails blasting out of rustic rocket ships," as he put it later. He went to work carving a whimsical 1950s science-fiction rocket lifting off on a dark and gnarled exhaust plume. He soon began to call himself a rustic rocket scientist, adding more wooden rockets as well as airplanes and planets to accompany his rustic furniture. NASA administrator Dan Goldin took such a shine to that first rustic rocket that he took it back with him from an air show at which Lindbergh displayed it, and installed it in his office in Washington.

Along with a new artistic sensibility, the X PRIZE also awakened in Lindbergh old fears about life in the public eye. Ultimately, he embraced the challenge and signed on as an X PRIZE spokesman and trustee—an effort that culminated in a solo flight of his own across the Atlantic in 2002 to commemorate the seventy-fifth anniversary of his grandfather's famous flight. Lindbergh didn't try to duplicate his grandfather's 1927 flight; he flew a modern craft with a global positioning system to guide him, and he was in constant communication with Gregg Maryniak, who acted as flight director from X PRIZE headquarters in the St. Louis Science Center. Lindbergh's purpose was to honor his grandfather's achievement while generating press for the X PRIZE along the way. But it had another, more profound effect on Lindbergh's psyche. "It was an amazing inner journey," he reflected afterwards, "pushing through some old family barriers and really setting myself free from the burden of the legacy. It allowed me to walk in Grandfather's footsteps without trying to fill his shoes."

Diamandis formally launched the X PRIZE with an outdoor press conference under the St. Louis Arch on the morning of May 18, 1996, two days shy of the sixty-ninth anniversary of Charles Lindbergh's flight. Among those joining him on the stage were Erik Lindbergh, *Apollo 11* moon walker Buzz Aldrin, and—talked into it by NASA officials friendly to Diamandis'

idea—NASA chief Dan Goldin, who proclaimed the X PRIZE a "noble venture." That night in St. Louis the X PRIZE founders held a black-tie dinner for contributors paying $500 a plate. During a meal of wild mushrooms and beef tenderloin, California aircraft designer Burt Rutan, famous for designing *Voyager*, the first airplane to fly around the world without landing or refueling, climbed up on the stage to make an announcement. "I have dreamt about making a homebuilt spacecraft ever since I've been doing homebuilt airplanes, for godsakes," he told the crowd. "But I have never been, myself, as creative as I have in the last couple of months, eyeballing this goddamn prize. And I'm not going to tell you what I've come up with because I want to win this thing." He grinned for a moment, letting that sink in. Then, "But I am going to tell you that I'm not the only one that's going to be creative." He predicted a rash of entrepreneurs trying to build private spaceships, "who are going to come after me like crazy."

He was right. Rutan was the first, but many more X PRIZE contestants soon joined in—twenty-six in all, from seven countries, before the prize was won. They included a retired NASA engineer working out of his garage on his pension, a fighter pilot turned down for the astronaut corps after making a wisecrack to one of his interviewers, a top video game programmer grown bored with coding, and a self-taught industrial designer from Toronto determined to build and fly the first Canadian spaceship.

The race was on.

2

Go!

The New Space Race Takes Off

"How about this one?"

Brian Feeney and I stood in a Home Depot in the northeast suburbs of Toronto, scrutinizing tape measures.

Feeney looked at the tape measure I pointed to and shook his head. "It doesn't have metric."

I suggested another one, but at more than $18, it was too expensive.

"How about that one?" said Feeney, pointing to a little red plastic job hanging next to the more expensive metal ones.

"This one? Are you sure?" It looked wimpy to me. Hardly something you'd use to build a spaceship. Then again, Home Depot hardly seemed the place to shop for spaceship supplies.

"That's the one," insisted Feeney. "Lock and load!"

The tape measure wasn't the only thing that didn't

seem as if it was meant to build a spaceship. Brian Feeney was a self-taught industrial designer by trade, forty-four years old when I caught up with him in October 2003. He had thinning hair and wore yellow-tinted glasses framed in black plastic, and had a silver earring in his left ear. He was a night owl, spending late nights at his local pubs, where he was known by all the regulars. But by day, he headed up the da Vinci Project, one of the better-publicized X PRIZE contestants. He talked about the project that had consumed his waking hours for seven years with a quiet intensity and a slight smirk, as though he knew perfectly well how crazy his plan to launch himself into space in, as he termed it, a "homemade rocket" sounded to other people.

Feeney's odyssey began in 1996, soon after the X PRIZE officially launched. He was living in Hong Kong at the time, sharing a flat with three other guys and working as a designer on contract with the Brita water filter company. His globe-trotting lifestyle shot holes in his marriage back in Toronto, and the marriage crashed and burned while he was in Hong Kong. So he stayed on, soon enjoying a life of nonstop partying. "Boy, did we play hard," he told me with a laugh. After his contract with Brita expired, Feeney found work for other companies, among other things designing gadgets for deaf people such as a vibrating alarm clock. But from that fateful day in 1996 when he read about the X PRIZE while browsing at his local newsstand, the idea of building a spaceship consumed him. He began sketching out ideas that very night, and soon afterwards he got on the phone with aerospace companies, getting feedback and sourcing supplies. It wasn't long before the high life in Hong Kong began to seem like a waste of time. Toronto, he decided, would be a better base of operations for his new project. Then, too, he'd begun to miss his two daughters back home. They were then nine and twelve years old, and both of them would come to enthusiastically support what Feeney eventually called the da Vinci Project.

When I arrived in Toronto at the start of a weekend, the only rental car I could find was a van. That was an annoyance at first—until I found out that Feeney needed to make a Home Depot run, and I was able to offer up the van for hauling supplies. Feeney hesitated, but I pressed him, and practical considerations overcame any reservations he might have about recruiting a reporter to work on his project; his little sedan was too small for what he had in mind. I had some vague notion that this wasn't proper journalistic protocol, that I should keep a professional distance from my subjects. But I *wanted* to get involved. And one of the major draws of an entrepreneurial spaceship project was that one *could* get involved.

After returning a defective power drill and loading our cart with assorted other tools and plywood sheets, Feeney and I headed to the checkout counter, where Feeney produced an envelope stuffed with cash. The money had been donated to the project by Sun Microsystems of Canada. Feeney wouldn't tell me just how much the big computer maker had given him, but it represented a major breakthrough. Although the project had successfully solicited in-kind donations of materials, office space, and labor for years, this was the first hard cash it had received to date, and it would allow Feeney to finally start construction on his spaceship. He grinned at me as he pulled the cash out of the envelope. "Rocket money!"

Feeney was but one of many entrepreneurs working on homemade spaceships at the turn of the twenty-first century. True to Rutan's prediction, Diamandis' X PRIZE vision sparked an explosion of such efforts, with varying degrees of feasibility, some better funded than others. At the time of my visit with Feeney, a retired NASA propulsion engineer named Jim Akkerman was working on a suborbital spaceship of his own in Houston, Texas, where he had been born, grown up, and worked all his adult life. A devout Christian, Akkerman had prayed for help in naming his fledgling rocket company, and the answer came back to him in short order: Advent Launch Services. "Advent" because it's the beginning of the word "adventure" and means "a new beginning," as Akkerman later explained.

A few of Akkerman's former colleagues helped him on his project as they found the time, but mostly Akkerman was on his own, machining his designs in a friend's machine shop in Houston near his home, and testing them in a rice field outside of town. That suited him just fine. After trying to buck the NASA bureaucracy for thirty-six years, he'd retired in 1999 (or "graduated," as he preferred to call it) to work on his own designs, and he welcomed the opportunity to keep his own counsel.

Akkerman's pet project while he was at NASA had been a rocket engine powered by liquid methane and liquid oxygen. Methane, insisted Akkerman, was a cheaper, more efficient fuel than the expensive liquid hydrogen favored by NASA for its space shuttle main engines. For starters, methane costs only pennies per pound to produce, versus several dollars per pound for liquid hydrogen. Seven times as dense as liquid hydrogen, it also packs more energy per cubic foot, requiring tanks a quarter or less the size of those required for hydrogen. With smaller, lighter tanks, a methane-powered spaceship would have less of itself to lift and could therefore carry a bigger payload.

The cost savings for methane became clear to Akkerman in 1992, when NASA managers at the Johnson Space Center, where he worked, put him in charge of a group studying the cost and safety benefits of switching from the current solid fuel to liquid fuel for the space shuttle's strap-on rocket boosters. Akkerman found that the fuel for the solid rocket boosters, or SRBs, cost a whopping $20 a pound. On top of that, the fuel cost an additional $20 a pound for transportation and final processing before it could be used. With about a million pounds of fuel required for a shuttle launch, the fuel costs for the boosters were, well, astronomical. It seemed like a no-brainer to Akkerman and his small group that, from the point of view of cost savings alone, NASA should make the switch to liquid fuel for the boosters. In fact, Akkerman recalled later with some bitterness, "we were making such a good case for a liquid booster that they terminated our study."

As Akkerman saw it, politics, not engineering or safety considerations, killed the liquid rocket booster study. NASA's Stennis Space Center in Mississippi was trying to advance its own design for a new and improved solid rocket booster called the advanced solid rocket motor (ASRM), and it had powerful allies in Washington. So, Akkerman and his colleagues got reassigned to other projects while the ASRM project went ahead. Expenditures on the ASRM soon spun out of control, and NASA axed that project in 1993. That left the space agency stuck with the same flawed SRBs that had doomed space shuttle *Challenger* in 1986 when hot gas blew past an O-ring that was supposed to contain it. The *Challenger* disaster made it all too clear that the solid-fueled rockets were a bad choice for the shuttle for reasons other than just their expense relative to liquid-fueled rockets; their inability to shut down in an emergency and their segmented construction requiring the use of O-rings also made them more dangerous.

Akkerman and the former members of his study group didn't give up on their methane engine design. They continued their discussions over the office coffeepot and in the lunchroom. These were some of NASA's finest and brightest, recruited to push the limits of space technology in the early 1960s during the run-up to the moon. They included, in addition to Akkerman, who had cut his teeth on Apollo moon mission hardware, Charlie Stamps, another Apollo-era veteran, who was skilled at navigating NASA's labyrinthine budgeting process, and Glenn Smith, who had been a project manager for space shuttle systems engineering during the program's early years. They and others continued their design work on the methane engine and added an orbital spaceship to it for good measure.

They developed a plan for using the Gulf of Mexico as their launching pad. A two-stage rocket would bob upright in the water like a buoy prior to launch. The booster and the orbital stage would return separately to water landings in the Gulf, ready to be refueled and launched again. The system did away with the needless complication of launch pads and runways, as Akkerman and his colleagues saw it. They were sure it would work and would be a cheaper and more reliable means of sending people into space than the space shuttle. So sure, in fact, that they continued to press NASA brass to give their project official sanctioning. But NASA wasn't interested.

So, working on his own time, Akkerman put together Advent Launch Services as an independent company and went shopping for backers. Again, no dice. Finally, "You know what?" he said, "I'm building this rocket myself." When the X PRIZE launched in 1996, the Advent team scaled back its plans and began work on a suborbital spaceship. Akkerman and his crew figured the $10 million prize would help fund the full-up orbital vehicle.

Akkerman's ultimate mission, like that of Diamandis and Maryniak, was to work toward making the world a better place to live in. His low-cost orbital ships would enable the construction of orbiting solar power stations that would beam down all the cheap energy the world could use, solving the simmering energy crisis. And he agreed with Diamandis and Maryniak that fostering space tourism was an important step toward the peaceful resolution of humanity's problems; it would help as many people as possible to see firsthand that all nations are dependent on the same closed ecosystem and that we would be limited by Earth's finite resources as long as we remained Earthbound. "Seeing the world from orbit is a life-changing experience," an astronaut friend had told Akkerman, and he believed it. Akkerman had done his best to get the big picture himself by buying himself an airplane when he graduated from engineering school in 1961, and then going to work for NASA. But ultimately he came to realize that he had to aim higher. Even the X PRIZE was just a small step along the way. "Our goal is not just to win $10 million," Akkerman told me in 2003. "If we can put things on orbit for under $100 a pound, we can make $10 million about once a week."

. . .

Back at the Toronto da Vinci Project workshop, not far from the Home Depot, the rest of the team had been busy. With the money from Sun only recently in hand, Feeney had only just been able to rent the workshop. It wasn't much—just a few hundred square feet with a makeshift loft in back of a scuba shop. Although it occupied one of dozens of identical blank box warehouses on the outskirts the city, the workshop was easy to find: a full-scale model of the da Vinci Project's spaceship, *Wildfire*, was parked out front. Bullet-shaped, painted the colors of fire, and sporting a sixteen-window cockpit inspired by the B–29 bomber, the mock-up had been built as a showpiece to attract sponsors. The workshop was so small that the mock-up had to be wheeled out each day to make room for the people working there.

Feeney's dream began when he was a boy growing up in Toronto's Leaside neighborhood. He remembered the exact moment with feverish clarity. He was six years old, at home sick, watching the family's black-and-white TV while his mother ironed in the same room. So powerful was the memory that even decades later Feeney could still smell his mother's iron as the words "Via Satellite" appeared on the screen and a Gemini spacecraft lifted off, carrying two American astronauts into orbit. From then on, Feeney read all he could get his hands on about rockets and space travel. When he was in the fourth grade, he made his first attempt at flight: he climbed up a tree with an umbrella and jumped. "For a split second it held," Feeney later recalled. "And then it went *foompf*" and flipped inside out, and the budding rocket scientist plunged to the ground, fortunately unharmed.

In the summer of 1969, even as Buzz Aldrin and Neil Armstrong and their colleagues were headed to the moon, ten-year-old Feeney and two of his school buddies were hard at work on a succession of increasingly powerful homemade rockets powered by a brew of saltpeter, sulfur, and charcoal concocted by Feeney in his parents' garage. They launched these Super Height Interplanetary Travel spaceships, as they called them, from their schoolyard until one day the last and most powerful of the line exploded on launch into a towering mushroom cloud. Feeney and his friends jumped on their bicycles and tore out of there before they could be caught.

Feeney chalked it up to experience and continued his tinkering in his garage lab, eventually constructing a homemade flame thrower. That experiment worked even better than he expected, and the first time he tried it he set a pile of boxes in a corner of the garage on fire. He reflected then that he should really point the thing out the open garage door. But once again, he underestimated the power of his new weapon, and the flame set the front tire of his grandfather's Rambler on fire on its way out. Fortunately, the flame thrower's fuel, an alcohol-and-water mix, burned itself out quickly, and Feeney was able to clean up the damage without his parents' being any the wiser.

In fact, Feeney's parents usually had little idea of what their son was up to in their garage. With four other kids to look after, they were happy just to have him out of their hair. "I was completely self-propelled," Feeney told me of his childhood. It also didn't hurt that he regularly won prizes at science fairs for his projects, including a homemade telescope for which he won first place, and a meticulously detailed pencil drawing of an Apollo command module of the type that took Armstrong, Aldrin, and their crewmate Mike Collins to the moon.

"Want to get your hands dirty?"

There were five of us in the da Vinci Project's new shop on Saturday morning: Feeney; me; Lorne Brandt, a forty-one-year-old physics teacher who had been volunteering for the project off and on for a couple of years; and a magazine-writer-and-photographer pair from Germany making a tour of X PRIZE competitors. Feeney had cut templates out of some foam core pieces

The da Vinci Project's spaceship displays sponsor logos at the 2005 X PRIZE Cup in Las Cruces, New Mexico. *(Photo: Michael Belfiore)*

we'd gotten at the Home Depot, and now it was time to trace the shapes on plywood and cut them out with a power saw. This was the opportunity I'd been waiting for, though I tried to sound professionally matter-of-fact as "Sure," I told Feeney.

"Do you have any experience working with tools?"

"Well, I took shop in high school," I offered.

Feeney chuckled. "That's more than a lot of us." He handed me a pair of goggles, and we got to work. He watched my inexpert handling of the power saw for a few minutes before taking it back and blasting through the plywood with quick precise cuts. "You're thinking about it too much," he shouted over the scream of the saw. "You have to just rip through it!" I got the hang of it after a few more tries and then managed to cut out a rib of plywood that, with eighteen other such ribs, would form the support, or "tool," for the Kevlar and other materials that would be built around it to form the nose of the spaceship. Feeney looked over my work appraisingly. Then, "Want to sign it?"

I hesitated. Would signing my name to the work of the da Vinci Project represent some kind of conflict of interest? The German journalist eyed me disapprovingly. "Go on, sign it," urged Feeney. "It's gonna end up in a friggin museum some day!" Screw it, I thought. I picked up the ballpoint pen I had used to trace out the rib on the plywood. But that wasn't good enough for Feeney. He wanted me to make it bold, with a thick permanent ink marker. My thoughts turned to the workers who had proudly signed the inside of the nose of the *Spirit of St. Louis.*

It was a measure of Feeney's effectiveness as a manager that he made everyone, including me, feel proud of their contributions to the da Vinci Project, no matter how small. It was a skill he'd honed during his tenure as CEO and founder of a profitable life-support systems company in the 1980s. Making a range of products, including breathing equipment for firefighters and rescue gear for miners, Feeney's company thrived on U.S. Department of Defense contracts and had him commuting twice a week between Toronto and company headquarters in Hartford, Connecticut. Then, in the late 1980s, recession hit. The defense contracts dried up, and the company's revenue plummeted. By 1990, Feeney was wiped out. He lost everything, including his house, his Mercedes 300, and his Porsche. Three months later, he landed on his feet with the freelance design job with Brita in Hong Kong.

Returning to Toronto in 1997 to start the da Vinci Project after six years in Hong Kong, Feeney had his work cut out for him. Securing financial backing for the project proved to be impossible, at least at first. Nevertheless, he finalized the basic design of his spaceship on his own and began building a team of volunteers. He also solicited donations of materials and in-kind support. In a happy coincidence, the da Vinci Polytechnic Institute donated office and meeting space near downtown Toronto. The project now had a place to call home.

41

To recruit his first volunteers, Feeney approached the University of Toronto's Institute for Aerospace Studies, where he found a ready supply of engineering students eager work on a cutting-edge aerospace project.

Underemployed engineers certainly were not in short supply in Canada, as I experienced firsthand when I went to pick up my rented van in a mall downtown. The young man behind the counter at the rental car company lit up when I told him I was on a visit to the da Vinci Project. Turns out, fresh out of engineering school himself, he was then working the only job he could find. Like many Canadians I talked to during the course of my visit, he spoke wistfully of the Avro Arrow, a supersonic fighter jet built in the 1950s by the A.V. Roe Aircraft company of Malton, Ontario, near Toronto. By all accounts ahead of its time, the plane advanced aerospace technology in several areas, including an innovative fly-by-wire system. But the Canadian government canceled the program in 1959 even as it got off the ground. Many Canadians cite this failure of vision, as they see it, as the genesis of Canada's frustrating propensity for ceding technological leadership to the United States.

Feeney's plan to build the first Canadian spaceship fired the imagination of many a starry-eyed engineer. The first volunteers came aboard the da Vinci Project early in 2000, and that June the da Vinci Project became Canada's first entrant in the X PRIZE competition. Also in 2000, Feeney gained one of his most powerful allies, Dr. Vladimir Kudriavtsev. A mechanical engineer specializing in computational fluid dynamics, the science of measuring and predicting airflow and its effects on an aircraft or speeding rocket, Kudriavtsev was chief engineer at a company called CFD Canada. Introduced to Feeney by a professor at the Institute for Aerospace Studies, Kudriavtsev quickly became the da Vinci Project's head of engineering. What the da Vinci Project lacked in money, it was now beginning to make up for in technical expertise from volunteer engineers looking for challenges they couldn't find in their day jobs.

John Carmack (left) and Peter Diamandis at the 2005 X PRIZE Cup.
(Photo: Michael Belfiore)

At the same time, another X PRIZE competitor, Armadillo Aerospace of Mesquite, Texas, faced just the opposite problem: plenty of money but a serious shortfall in the engineering department. But the team's leader, John Carmack, was determined to get ahead of the learning curve, *fast*. In fact, he lived for just this kind of engineering challenge. And his innate thirst for technical knowledge had already made him a fortune.

I first encountered Carmack's work in December 1993 while visiting a friend of a friend in Minneapolis, where I then lived. I've long since forgotten the name of my friend's friend, but I'll never forget what happened when he booted up the minitower PC sitting on his living room floor. The screen on the desk above exploded to life with images of another world: a dark nasty world, with demons lurking around the corners of scum-encrusted corridors. Pits of poison slime bubbled in vast rooms with vaulted ceilings and overflowed from rusting toxic waste barrels. Visible through high arched windows, storm clouds billowed overhead, driven through blood-red skies on a devil wind. But what made

43

the experience so memorable was that *I was there*. With my left hand resting on the keyboard and my right on the mouse, I could control what I saw on the computer screen. I could look left, right, forward, back. Two gloved hands gripped a .45 automatic in front of me, pointing down the corridors. My trigger finger twitched over the left mouse button, and the pistol jerked upward in recoil and spat flame and bullets. Around the next corner, a demon leaped on me with a horrifying growl. Startled, I emptied the clip of my pistol into the beast, and then, left with no other option, I began slugging it out with my fists. The first demon went down, but more piled on until the world turned red and tilted away and I died with a scream.

It was the first time I'd actually been *scared* by a game. By countless horror movies, yes, but never by a game. It felt so real, so visceral. Those growls. That awful scream. You never knew what lurked around the next corner, just as in a good monster movie. But instead of having to sit and sweat it out, you could fight for your life, and maybe even blow those bastards away with pistols, fists, a chain saw, shotguns, rocket launchers— whatever you could find stashed away in those gloomy corridors, behind hidden doors in secret rooms. If you were *very* lucky, you could score yourself the ultimate weapon, the BFG9000 (where "B" stood for "Big" and "G" for "Gun"), capable of mowing down whole armies of beasts from Hell.

For millions of video gamers, *Doom* was unlike any video game ever made. More realistic and more addictive than anything that had come before, *Doom* set the video game world on fire that December of 1993. The day of its release on December 10 via file transfer protocol, it brought its host computer network at the University of Wisconsin to its knees under the onslaught of 10,000 gamers. Within a day, the game's developer, id Software, of which Carmack was part owner, was pulling in $100,000 a day in sales, with a hundred times the number of purchasers downloading free shareware versions of the game.

Much of the game's success was due to the fact that it ran

on an ordinary personal computer. Until then, the best games were relegated to quarter-eating arcade machines. Now everyone could own a piece of the best gaming action in the world. I looked at my friend's machine on the floor with new eyes. It seemed to glow now, as though imbued with magical powers. Could I too get a machine capable of this magic? Most assuredly yes, as it turned out. It was the programming genius of John Carmack that made it so.

A self-taught programmer just twenty-three years old when *Doom* launched a video game revolution, Carmack lived, ate, and breathed code. Taking the work of game designer and fellow id co-owner John Romero as his blueprint, Carmack programmed, tweaked, hacked, and finessed his way to a series of elegant solutions that put spurs to the processing powers of ordinary IBM-compatible personal computers of the early 1990s and enabled them to create immersive alternate realities. Carmack's over-reaching mission in all this was the quest for speed. Faster graphics rendering, faster action. When in short order he bumped up against the limitations of his chosen platform, the IBM PC, he found clever ways of working around them. He used an advanced programming technique called binary space partitioning to enable the machine to render the graphics of the game's environment only in the split second when they were needed as the player looked around, instead of bogging down in rendering the entire scene at once. Previously this technology had been used just for rendering individual objects on a computer screen, but in a major breakthrough in video game programming, Carmack applied it to rendering dynamic scenes as gamers moved through them.

He explored the concept of speed in the real world, too. With the help of Dallas hot-rodder Bob Norwood, he souped up a series of $70,000 Ferrari sports cars with turbochargers, transforming them into "1,000-horsepower monsters," as he later put it to me, and pushing them past two hundred miles per hour.

Seven years and half a dozen iterations of *Doom* and its *Quake*

sequel later, Carmack found himself running out of challenges. He'd reached the pinnacle of his profession, rewriting its rules in the process, and he was growing bored. He'd mastered his craft, and if he was lucky, he'd pick up a new skill just once every few months. It wasn't enough to keep his mind engaged. So he began to look for other creative outlets.

Unlike so many of the other X PRIZE competitors, Carmack hadn't always been obsessed with space. Though he had "had a standard geek childhood in that it involved model rockets and science fiction and stuff like that," he "really didn't think about space for over ten years while I was building my software business." But after experimenting with increasingly more powerful model rockets, Carmack realized this was a pursuit worthy of serious effort; here was an area of vast new technical challenges.

In 2000, Carmack jumped in feet first, exchanging e-mails with the president of the Dallas Area Rocket Society to find a team of fellow enthusiasts who would help him build a manned rocket. Not necessarily a spaceship, but a "vertical dragster." Good enough for amateur rocketeers Russ Blink, Phil Eaton, and Neil Milburn, who answered the call. Bob Norwood joined the team as well and donated his Ferarri shop as an occasional place to work (along with the back of Blink's pager company warehouse), and Armadillo Aerospace was born.

Carmack and his crew got together Tuesday nights and Saturdays to design and then machine from scratch small hydrogenperoxide–powered, pressure-fed rocket engines. Carmack bought a hundred acres outside of town, and there Armadillo's designs took flight on powerful homebuilt model rockets. Once again, Carmack was in his element, as in the old days of programming *Doom*, "soaking up huge amounts of information all the time." Carmack learned all he could of rocket design from NASA technical reports out of the 1960s and 1970s ("the recent ones are just useless; they're studies about the studies that have been done before"), and then, just as he had done in the realm of software design, he sought to improve on the state of the art.

"Rocket science," he concluded, "has just been mythologized out of all proportion to its true difficulty; it's not as hard as people think it is." None of the team had any aerospace experience beyond model rocketry, but that didn't bother Carmack. "We learn what we need to as we go," he told me, "and that's been my primary motivator for doing all of this."

In fact, signing up for the X PRIZE competition came almost as an afterthought for Armadillo, and not until October 2002, when it became clear to Carmack that the X PRIZE organization actually had the funds with which to award the prize money. By then, Armadillo was building increasingly advanced rockets that took off and landed on their tails like 1950s science fiction spaceships. These vertical takeoff/vertical landing vehicles were small; at their most powerful, they lifted just one person seated in an open chair between propellant tanks a few feet into the air. But they allowed the team to perfect techniques that Carmack felt sure he could easily scale up to full-size spaceships.

True to his background as a software engineer, Carmack created sophisticated computer controls for his rockets. "Our core strategy for everything was that we were going to attempt to do as much as possible with software and electronics," Carmack told me. "Any place where we can minimize or remove a mechanical system by using more sophisticated electronics, that's what we were going to do." From the beginning, each Armadillo rocket carried an onboard computer that correlated data from gyroscopes and accelerometers to precisely plot the rocket's trajectory and location (later rockets included global positioning systems). An ordinary wireless router on the rocket networked with Carmack's laptop on the ground to let Carmack control the rocket in real time with a video gamer's joystick. Moving the joystick commanded multiple rocket engines on the vehicle to fire in precise sequence to launch the rocket; make it hover at a desired altitude, even in the face of a stiff wind; to translate sideways; or, with a press of a joystick button, to land automatically. Armadillo's rockets were doing all that and more by the time Ar-

Armadillo Aerospace's Lunar Lander Challenge vehicle takes shape in 2006. *(Courtesy of Armadillo Aerospace)*

madillo signed on as an X PRIZE competitor. It didn't seem like too much of a stretch for the members of the team to think they had what it took to go all the way to space.

Armadillo's full-size spaceship would launch vertically like its smaller predecessors. After reaching space, it would nose over to point back to Earth. Parachutes would slow it down after re-entry, and a crushable nose cone would collapse on impact with the ground like an automobile bumper, further protecting the crew. After replacing the nose cone (at just thirty-seven pounds, it came well within the 10 percent of vehicle weight allowed for expendable parts) and refueling the vehicle, the crew would be ready to fly the ship again. As the only member of Armadillo Aerospace with anything approaching flying experience, albeit only as a skydiver, Russ Blink became the team's designated astronaut. Since the X PRIZE rules allowed the winning team's pilot to fly with the weight equivalent of two people in lieu of actual passengers, Blink would make the trip alone. Carmack himself had little actual interest in traveling into space. "In truth," he told me, "the motivation for me is more grappling with the challenging problem."

By October 2003, team Armadillo Aerospace had drop-tested a full-size crew capsule mock-up from a helicopter over Carmack's hundred acres, successfully demonstrating parachute deployment and the effectiveness of the crushable nose cone. They'd outfitted Blink with a Russian spacesuit purchased on eBay that, while definitely used, managed to hold pressure after some patching, and they'd completed much of the spaceship. Carmack expressed confidence that they'd get it in the air by year's end and be ready for X PRIZE flights by the end of 2004. Carmack's bill for all this experimenting and rocket-building so far: a million dollars. "Which is a lot less than I spent on Ferraris," he told me. He didn't think he'd have to spend any more than a total of $2 million winning the X PRIZE. Which would give him a nice, tidy little profit. Not that he was unduly focused on profits. "I'm not really motivated as an entrepreneur so much as an engineer." This attitude had served him well during his rise in the computer gaming world, and he saw no reason to change it now.

"*This is* the *Tiger Shark*," Feeney told me. "If Ferrari were to build a rocket, it would look like this." The workday was over, and now Feeney and I were hanging out in Feeney's apartment not far from the Toronto neighborhood where he had grown up. We were checking out a pre-*Wildfire* space capsule design on a fresh-out-of-the-box twenty-four-inch LCD monitor donated by Sun Microsystems. Like the current *Wildfire* design, the *Tiger Shark* had sixteen cockpit windows, but it sported twin rocket engines in the rear instead of *Wildfire*'s single engine. No doubt about it, it was cool. As Feeney said, it looked like a sports rocket ship.

Feeney's apartment was surprisingly cool as well; it looked like nothing so much as a set from an Ikea catalog from the future, a perfect blend of modern form and comfortable function. I couldn't help marveling at how utterly efficient Feeney had been at maximizing every square inch of space—so much so that the tiny one-room apartment actually felt spacious. Feeney's work area, where we sat, occupied one corner of the room, his bed the opposite corner. The bed was screened from the rest of the apartment by curtains hung by discreet cabling from the ceiling. A row of seats from a Dash–8 Canadian-built airliner between the bed and the work area defined an inviting-looking "living room." A TV and sound system occupied a long low table of Feeney's design, which featured a sheet-metal cylinder, a pyramid, and a cube as supports. Another cube, this one of black plastic, looked right at home on the table among the stereo components. In fact, it was a humidifier Feeney had designed for Brita. Against the wall across from the table was a couch, and to its left, a space-saving set of shelves held together by cables under tension. "If you were to cut those," said Feeney, "the whole thing would go 'sproing!'"

If Feeney's apartment was any indication, Feeney's spaceship would be the picture of efficiency and thoughtful design, all of its components perfectly balanced under tension. Feeney certainly had ample motivation for making it so, and getting it right the first time, too, for there would be no unmanned test flights of the vehicle. Feeney planned to be aboard right from the beginning.

That first flight, although it was to be a full-up run to space with Feeney aboard, would not be an X PRIZE flight. It would be a test flight, with the first X PRIZE flight to follow after any bugs that cropped up were ironed out. But it would be a very public test, with full media coverage. "It's so damned expensive to do this," Feeney explained, "we simply cannot afford to have a private test. If you do all this, and something goes wrong and something doesn't work, well okay, you just did it in Technicolor in front of everyone. But if something goes wrong and you did it privately, you're just as set back. . . . So, hey, go for it."

Wildfire's odyssey was to begin a week before launch with an overland journey from da Vinci Project headquarters in Toronto to the tiny town of Kindersley, Saskatchewan, two thousand miles away, chosen for its remote location and its surrounding wide-open spaces. The da Vinci team would take over the airport and its single runway with the town's blessing (shortly after the da Vinci Project selected it as its launch site, the town enthusiastically began calling itself Cape Kindersley).

The countdown would begin at T minus three days. If the weather still looked good twenty-four hours before launch, the team would announce a provisional "go" and begin releasing weather balloons every four hours to get the most precise fix possible on the prevailing winds at every altitude. Conditions would have to be exactly right, or the launch would have to be scrubbed; as Feeney was fond of pointing out, "space has no sense of humor." Especially when it comes to using the world's largest reusable helium balloon as the first stage of a manned rocket ship.

The twenty-five-story balloon would take two hours to fill with helium. Stretched across the ground when first attached to *Wildfire*, the balloon would rise into the air as it expanded. *Wildfire*, by then fully fueled itself, would be driven on a mobile launch platform to launch position beneath the balloon.

When fully filled, the balloon strains at the end of its tether seven hundred feet above the spaceship, big as an office tower, delicate as a dandelion puff. At T minus ninety minutes, Feeney, wearing a modified scuba diver's dry suit in place of a spacesuit, climbs into the capsule. Feeney and his team have loaded the capsule with the required ballast to stand for the weight of two additional people—along with life-sized Homer and Bart Simpson blow-up dolls in the two passenger seats for good measure. Space may not have a sense of humor, but Feeney doesn't let that dampen his.

An hour later, the ground team closes the hatch, and Feeney activates *Wildfire*'s life support system. The system, based on designs Feeney developed for his life support system company in the 1980s, is an open loop system, meaning that air expelled from Feeney's lungs is allowed to leak out of the capsule at the rate of forty liters per minute—the same rate at which compressed air is drawn from the ship's carbon fiber air tanks.

After the final checks have been made, the ground crew releases *Wildfire*, angled at 15 degrees from the vertical, from its launch platform, and the balloon pulls it skyward. According to Feeney, the team will have "launched a rocket that's like the slowest launch in history. It's kind of like, 'there it goes . . . we can still see it.' Ten minutes later, 'we can still see it.'" In fact, *Wildfire* won't reach launch altitude for a couple of hours. Which will allow plenty of time for journalists to interview Feeney along the way via satellite. Cameras mounted inside and outside the capsule will also send live images to the ground.

At 80,000 feet, Feeney separates *Wildfire* from the balloon and lights the synthetic-rubber-and-nitrous-oxide-powered rocket motor with a *whoosh* barely audible in the thin atmosphere.

Seventy-five seconds later, at a peak acceleration of 4 g's, the engine cuts off, the capsule separates from the propulsion section, and Feeney coasts the rest of the way to space, weightless.

Two minutes later, when he hits the 100-kilometer mark, he gets on the radio to the reporters: "We got to space! We're in space! We're talking to you from space!" He doesn't have much time to enjoy the view, however. Almost as soon as he arrives, he reaches apogee and starts back down. Now he must fire the reaction control system, or RCS, to properly orient the capsule so that the blunt end with its attached heat shield faces his flight path.

The capsule decelerates rapidly when it hits the atmosphere, and Feeney goes from 0 to 7 g's in about fifteen seconds. In fact, it may be enough to turn some of the air around the speeding capsule into a plasma that could block radio waves from entering or leaving the craft. Feeney won't know until he gets there, but he's prepared either way. "What I thought I'd do," Feeney told me, "if there is no blackout at all, and they're talking to me, I'm not going to answer anyway. I'm going to let them think there was a blackout. Why?" He grinned, ever the joker. "Why not?"

After the capsule has decelerated and descended sufficiently, the parachute pops out. He has no control of the trajectory the capsule takes on the way down. "We're not using a parafoil; we basically concluded it wasn't necessary. Now, if we land on someone's car, house, or barn, then I'll have a different opinion." Also landing on parachutes are the propulsion section of the spaceship and the balloon, which has released its helium before popping its parachute. All three sections are tracked by mission control at Kindersley Airport, with volunteers in chase planes and Jeeps in hot pursuit. The entire flight, from launch to landing, lasts about two and a half hours, with most of that time spent reaching 80,000 feet under the balloon.

Even one successful flight would accomplish what Feeney described as the da Vinci Project's main objective—to show that a private citizen, even one with limited means and no special training, could reach space without government backing, and to inspire others to do the same. "It's not about technology," said Feeney. "It's not even about commercialism right away, or private spaceflight to make money. It's simply about showing that it can be done and break down the barriers so other people with better ideas and more capital can step out there and say, 'Yeah, they showed the way, I'm going to follow them.'"

Twenty-six teams. Each with a different proposal for achieving the same goal—sending three people into space without help from any government and winning $10 million. They worked in garages, in donated machine shops, in fallow fields and cow pastures. Some of them had relevant aerospace experience, some were doggedly determined to learn what they needed to know in time to win the prize. A few had millions of dollars in backing, though most toiled on whatever they could scrape together just to keep their dreams alight each day. And in the end, only one could win the X PRIZE.

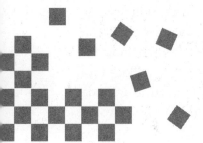

3

THE HOMEBUILT SPACESHIP

Burt Rutan's Big Idea

The epiphany hit in the middle of the night. Suddenly Burt Rutan was wide awake, staring into the blackness of his bedroom ceiling. Out of the blackness came a spaceship in free fall, heading toward impact with Earth's atmosphere. "I've got it," he said aloud. "I've got it, I've got it!" Beside him, his wife, Tonya, stirred and propped open an eye. "Hnmm?"

"I know how to do this!" Rutan scrambled from bed, snapped on the light, and grabbed paper and pen. He explained as he sketched rapidly, before the vision could fade. The craft's twin tail booms would pivot upward, creating a high-drag configuration, just like the feathers of a badminton shuttlecock. It would slow the ship quickly as the dynamic forces created by the thickening atmosphere increased. They would also orient the ship belly down, keeping

the stress and friction-induced heating caused by reentry to a minimum—all without any input at all from the pilot. Then, when the ship had descended low enough in the atmosphere and had slowed to subsonic speed—low enough and slow enough for controlled flight—the tail would pivot back down and lock into place. The pilot would then take hold of the controls to glide it home. "Carefree reentry," Rutan called it. He knew it would work. And it was the key to building a spaceship safe and affordable enough for the public to fly—the solution to a problem that had been plaguing him for years.

Deeply tanned by the desert sun and sporting old-fashioned muttonchop sideburns, Rutan would look just as at home on a cattle ranch as a runway. A gaze focused somewhere over the horizon adds to that impression, but it's a different kind of open range that keeps his mind whirling with possibilities, sometimes at the expense of those around him. Once while he was in the middle of a golf game with one of the test pilots at Scaled Composites, the company he heads in Mojave, California, a thought struck him: a possible solution to an engineering problem he had been turning over in his mind. Without a word, Rutan began walking, following that thought right off the golf course, into the parking lot, into his car, and back to his workstation, leaving his befuddled golfing companion standing alone on the course.

Rutan and his older brother, Dick, got bit hard by the flying bug while they were growing up in the 1950s in the little town of Dinuba, California, where their father was a dentist. The brothers lived in a converted shed in the back of the garage. As they talked into the night their dreams seemed to boil up out of the mind of one and infuse those of the other. Though they both loved flying machines, they took off on different headings. Dick devoted most of his energies to learning to fly anything with wings and a propeller or a jet, while Burt spent his boyhood inventing and building model airplanes. He didn't bother with the store-bought models his brother favored but built his own designs, soon turning the little shed bedroom into a workshop full of model airplane wings and fuselages, tails, and propellers like the dissected parts of enormous prehistoric insects.

Both brothers flew solo in airplanes as soon as they legally could, at sixteen. Burt Rutan went on to earn an aeronautical engineering degree from California Polytechnic in 1965, and he immediately went to work at Edwards Air Force Base near Palmdale, California, as a civilian flight test engineer. Edwards seemed the place to be for anyone who wanted to work at the cutting edge of aeronautical design. That was where, in 1947, Chuck Yeager became the first person to fly faster than sound. It was where rocket planes, fighter jets, and hot new airplanes of all descriptions made their first flights, where the best pilots in the world fought to master the world's fastest machines.

While Dick joined the Air Force and flew combat missions in the Vietnam War, the younger Rutan supported the war at home at Edwards with flight tests of the F–4 Phantom, a fighter-bomber with an unfortunate tendency to stall (lose lift), and spin out of control when overloaded with fuel and munitions. Green though he was, Burt Rutan attacked the problem with his characteristic single-minded focus, flying back seat to pilots who deliberately spun the airplanes to study the plane's flight characteristics. He wrote up flight procedures for recovering from spins, and for avoiding them in the first place, literally writing the rules on how to tame this particular beast. His work earned him the Air Medal, a high honor for a civilian.

Still, Rutan couldn't be fully content working and flying other people's designs. "I found even though I had a very exciting job," he said later, "that I had even more fun at home developing my own airplane." For him, nothing could beat the solitary satisfaction of transforming drawings into three-dimensional parts of plywood, glue, and fabric with his own two hands. And with his new-found engineering skills, he faced an irresistible challenge—designing and building a plane that he could actually climb aboard and fly.

To launch himself into the air in his own designed-and-built flying machine—now that would be the ultimate ride. Even better: what if he could somehow build not just any plane, but something that approached the agility and responsiveness of a fighter jet? For inspiration he turned to a Swedish fighter called the Viggen, built by Saab starting in 1962. That plane distinguished itself with extremely short takeoffs and landings, helped by the extra lift provided by a pair of stubby secondary wings on the plane's nose called a canard. The idea was that the plane could be rolled out of a bomb shelter in the event of nuclear war and launched from any available stretch of road. In a characteristically wild creative leap, Rutan saw that he could reduce the size of the Viggen design and create a sort of scale model of its handling characteristics in a smaller and slower propeller-driven plane. The effect would be to give the pilot of a homebuilt plane some of the adrenaline rush of flying a fighter jet.

The canard, a pusher propeller in the back, the lightweight design that could be built by a dedicated amateur working alone or with only a little help—all were characteristics that were to influence the dozens of Rutan designs that were to follow. Not to mention the "wow" factor that came from the idea that you could build a high-performance aircraft in your spare time in your garage. "I wanted an aircraft for myself," Rutan said later, and he might as well have been speaking for the legions of home builders who were to become fans of his later designs, "as close to a modern fighter as possible. . . . I wanted a big stick, an array of buttons, a high rate of roll—a real 'macho machine'—where I'd really feel like I was flying a century-series fighter." Such an outrageous ambition might have sounded mad to anyone who didn't know Rutan well. But it turned out to be entirely doable for the man who was shortly to turn the homebuilt airplane industry upside down.

In 1968, Rutan went to work building his plywood VariViggen, as he called it, in his Lancaster, California, garage. The craft's nineteen-foot wingspan and nineteen-foot length made it a tight squeeze in his garage, but he spent most of his off hours there, weekends and late nights, while his first wife, Judy, held a light for him to work by in a desperate attempt to spend more time with him. In 1970, Judy became the first of three wives to wipe out on Burt's obsession with building airplanes. She left him after seven years of marriage, taking their two kids with her. "There was no question in my mind which I wanted to keep," Rutan later said of his wife and of his greater love taking shape in the garage. There was no question, either of becoming further entrenched in the burgeoning government bureaucracy that was Edwards Air Force Base. He wanted to build and fly airplanes, not push papers, so as the VariViggen neared completion he quit the government for the private sector and never looked back.

Rutan rolled out the VariViggen for its first test flights in 1972 and then moved the plane, himself, and his new wife, Carolyn, to Newton, Kansas, where he put in two years working for light aircraft designer Jim Bede. Working for a designer of home-built airplanes was closer to Burt's heart, but he still wanted to be his own boss. So he also laid plans to strike off on his own with a company to be called Rutan Aircraft Factory (RAF), or simply "Raff." In 1974, Rutan moved back to the Mojave Desert, where he set up shop at the then nearly deserted Mojave Airport. The place had started life as a training base for Navy and Marine pilots during the World War II and now was little more than a collection of empty windswept roads, metal hangars, and a couple of runways. With hundreds of square miles of empty desert all around and miles and miles of clear blue sky overhead for all but a few days of the year, it was perfect for building and flying experimental airplanes.

Contrary to its name, RAF was not so much a factory as a small flight research center. Rutan built his own prototypes to test his designs, but his real product was the plans he sold to home builders all over the country, starting in 1976. The general aviation industry, whose customers were dominated by private amateur pilots, was in a bad way in the 1970s. After a boom in the 1940s and 1950s, the design of new airplanes had stalled in the

1960s because too many Sunday fliers with too few flying hours had "augered in," to use the aviator's euphemism for smashing to Earth at the speed of flight, and their surviving family members had sued the planes' manufacturers. To protect themselves, companies like Cessna and Beech simply stopped designing new airplanes for the private market. However, those with the time, inclination, and basic skills could still build their own new airplanes with plans and kits they could buy from independent designers.

Those independent designers could keep their liability to a minimum if they supplied less than half of the finished airplanes to consumers because the Federal Aviation Administration (FAA) considered each homebuilt plane a completely separate airplane type. No matter that dozens of these planes might be built from the same plans or even kits constituting as much as 49 percent of the finished airplane; each builder was in effect his or her own manufacturer, with materials and methods that might differ markedly from one another. Thus, each pilot flew at his or her own risk, leaving designers free of the heavier liabilities faced by the big manufacturers.

In the latter half of the 1970s, Rutan's star rose quickly in the homebuilt market with designs that were kind to the relatively inexperienced pilots who might fly them and that, best of all, were relatively easy to build using ordinary materials. Rutan was inspired by the fiberglass glider planes being repaired in a shop at the Mojave Airport. "I spent hours drooling over the smooth, contoured, efficient glass composite European sailplanes," Rutan said later. He saw no reason why home builders couldn't work with the same materials. The sailplanes were made by laying fiberglass over molds—too complex a process for homebuilders. So Rutan came up with a way of laying up fiberglass over foam that could be cut into shape by hand.

Composite structures made of layers of fiberglass cloth glued onto shaped foam cores were no more expensive than metal, much lighter, and far easier to work with. If a piece didn't turn out right, a builder could easily reshape it, add to it, or remove it entirely. In this way, building an airplane could become much like building a boat—certainly a reasonable-seeming project for a dedicated amateur. "This method is light, strong, requires no particular skills or tools," Rutan wrote in a newsletter to home builders, "and best of all, can be done in about one-quarter of the man-hours required to build the metal wing." The cloth made of glass fibers all aligned in the same direction and impregnat-ed with the epoxy resin glue that Rutan found worked best for his new construction method wasn't easy to obtain in the small quantities needed by each builder. So Rutan imported it from Eu-rope, where it was more commonly available, and then sold kits for building aircraft parts to his customers. RAF plans and kits soon became *the* way for amateurs to build airplanes.

"Both in its unorthodox construction techniques and its flight qualities there is a fascinating potential for a less expen-sive, more forgiving flying machine than we have coming off any aircraft production line today," *Popular Science* aviation editor Ben Kocivar raved in a 1978 cover story about Rutan's VariEze design. Riding back seat on a test fight with Rutan, Kocivar marveled that you couldn't stall this airplane if you tried. "It's an uncanny feeling," he reported. A leading cause of crashes for small air-planes was stalls, or loss of lift, when the pilot slowed into a turn on the final approach to the runway. Rutan's canard-and-pusher design prevented stalls, even at low speed, by imparting extra lift and keeping prop wash—which might disturb the airflow over the wings—safely behind the airplane.

At the time of Kocivar's visit to RAF, Rutan was selling plans for the VariEze for $191. These included instructions for building the wings and fuselage by gluing unidirectional fiberglass cloth in crossed layers (for maximum strength) over urethane foam cores that builders could cut to the right shapes themselves—the technique Rutan learned from his study of European sailplanes at the Mojave repair shop.

RAF's plans also included recommendations from Rutan on

suppliers for all materials and came with Rutan's hand-typed and photocopied newsletter sharing building tips and ideas. Time needed to build a VariEze: 600 to 1,000 man-hours. Cost of materials and off-the-shelf parts: $5,000 to $9,000, depending on how many of those parts you bought prebuilt, and whether you bought your engine new or used. The low end of that range was only a couple of hundred dollars more than the price of the Volkswagen Rabbit reviewed in *Popular Science* immediately following Kocivar's article. Build a VariEze instead of buying the BMW 320i sports car reviewed on page 30 (BMW's cheapest, according to *PopSci*'s editors), and you'd have a couple of thousand dollars to spare. No wonder home builders swamped RAF with orders—some 4,500 of them by the end of the 1970s.

In fact, by 1978, Rutan and Carolyn couldn't keep up with the demand for plans and technical support. Rutan found himself spending more of his time on the phone with builders, giving them advice and helping them when they got stuck as well as answering their questions by mail, than designing and building new airplanes. He needed help if he wanted to rise above the flood of paperwork, so when home builders Mike and Sally Melvill flew into Mojave in their new VariViggen to show it off to Rutan, Rutan signed them on as his first full-time employees. Mike went to work handling tech support and building and flight testing prototypes, and Sally became RAF's bookkeeper. By then, Mike recalled later, Rutan "was selling so many VariEze plans it wasn't even funny."

Still, it was soon to get a lot more busy at RAF. Rutan had Mike Melvill build the prototype for the next-generation VariEze, called the Long-EZ. With much more wing area, this plane could carry a lot more weight, including the starter, alternator, and full electrical system that the VariEze lacked, along with more fuel. From the start, said Melvill, "it was a much more popular airplane." In fact, Long-EZ kits and plans sold twice as well as those for the already successful VariEze. "We stopped selling VariEze kits and plans at that point because the Long-EZ is a much better airplane," said Melvill. "It flies slower, lands slower. It's a safer airplane, so it was the most popular one he did." Melvill himself liked the airplane so much that he built one of his own, and he was still flying it regularly twenty-five years later.

But even as Rutan's airplane designs established an enduring presence all over the country at air shows, in garages, and even in the movies (with a cameo by Rutan flying his VariViggen in 1975's *Death Race 2000*), Rutan's second marriage became another casualty of his single-minded focus; Carolyn bailed out in 1979.

As Rutan's reputation as an airplane designer with few equals grew, he began to attract attention from the big boys of aviation. For instance, Fairchild Republic. When this defense contractor needed flight test data, and fast, for a new jet trainer it was to build for the Air Force, it hired RAF. RAF built a 62-percent-scale model of the proposed trainer out of composites, and flight tested that. Most such preliminary testing of new aircraft designs was done in wind tunnels using models the size of hobby store kits. But this new method of checking out an aircraft design—manned flight in a subscale version—yielded better data than a wind tunnel and was much faster than building and testing a full-up airplane. Suddenly Rutan had a promising new market for his company.

A contract to build and flight-test an 85-percent-scale model of a business jet called the Starship designed by Rutan for Beech Aircraft confirmed that he really had tapped into a rich vein, and that set the course of his career for the next two decades. In 1982, he founded Scaled Composites, a company dedicated to this kind of work, next door to RAF at Mojave Airport. "We couldn't do it at RAF," Mike Melvill recalled later of the Starship contract. "We didn't have a big enough building. We didn't have any employees. So we started Scaled Composites just to get that plane built."

"Scaled" because the new company would build scale models of bigger airplanes. "Composites" because those airplanes would be of the same composite construction that had made Rutan's homebuilt designs so successful. But these homebuilts would be for the United States government and major aerospace companies looking for ways to save time, to reduce the weight of designs that would otherwise be of metal, and most of all to save money. In 1985, RAF shut its doors and moved into a back corner of Scaled Composites, where Melvill continued to provide technical help to home builders and airplane owners into the 2000s.

Burt Rutan, already famous in aviation circles, became known to the wider public with the record-breaking 1986 nonstop around-the-world flight of *Voyager*, a spindly carbon-and-glass-fiber airplane designed by Rutan and flown by Dick Rutan and his partner Jeana Yeager (no relation to Chuck). By then, Scaled Composites was designing, building, and flight testing an average of one completely new aircraft every year. *Voyager* now hangs in the National Air and Space Museum in Washington, D.C., along with the Wright Flyer, *The Spirit of St. Louis*, and of course the world's first homebuilt spaceship.

As early as 1994, Burt Rutan had been sketching ideas for a Scaled Composites craft that could go faster than the speed of sound and leave the atmosphere. In other words, he wanted to build a spaceship. More than that, he wanted to construct a *homebuilt* spaceship. Still infused with that wild-haired spirit of sheer inventive adventure that had let him dare to consider building a fighter plane in his garage, he wanted to take his homebuilding techniques to the ultimate extreme. Supersonic. Exoatmospheric. Nothing could beat a rush like that.

Could it could be done? At Edwards in the 1960s he had worked beside engineers for the X–15, a rocket-powered airplane that was launched from a B–52 bomber to reach seven times the speed of sound and top fifty miles in altitude, the then widely recognized boundary of space. This craft wasn't fast enough to actually go into orbit—for that it would need to go twenty-five times the speed of sound. Even so, the technical challenges posed by this suborbital ship were legion.

The force of the wind created by the ship's supersonic passage through the air rendered ordinary hand-operated flight controls immovable past the transonic boundary. Once out of the atmosphere, the ship's controls could be operated again, but they would do the pilot absolutely no good; with no air to dig into, the ship's ailerons, rudders, and other control surfaces could flip and flap all they wanted with no effect whatsoever. Alternative space-only controls, called a reaction control system, or RCS, had to be used to orient the craft as it sailed through space. The RCS used little rocket motors firing in response to control stick inputs to orient the craft in the three axes of pitch, yaw, and roll. Of course *those* controls became useless when the craft reentered the atmosphere.

If the ship hit the atmosphere the wrong way, with a wing tilted downward, say, or even worse upside down, it could easily be torn apart by the aerodynamic pressure of the supersonic wind exerting itself in ways the ship wasn't designed to tolerate. An experienced test pilot had to use every bit of skill he could muster to get himself and his ship home safely, to make sure the loads were distributed properly along the length of the ship's airframe, and not to take them broadside. Right at this most critical phase of its flight, the ship would have to seamlessly make the transition from RCS to atmospheric flight controls.

Rutan was at Edwards in November 1967 when one of the X–15's test pilots, Mike Adams, mistook a roll for a yaw because of a quirk in the X–15's attitude (orientation) display. The display required the pilot to read it differently depending on whether he was heading up or down. Adams reentered the atmosphere sideways, and the ship went into a spin while traveling at hypersonic speed (greater than Mach 5). As the ship slammed sideways into the hammer blow of the wind it simply flew apart. Adams didn't have a chance.

That accident put the fear of God into Rutan as far as the idea of a winged spaceship was concerned. The X–15 absolutely depended on automatic control systems to help the pilot fly the ship on the extremely precise course it needed to survive reentry, and even with them, as Adams's accident showed, successful reentry was an iffy proposition. The X–15's electronic controls held the airplane at the attitude, or orientation, the pilot selected as he reentered the atmosphere. They also damped the motions of the plane's control surfaces caused by supersonic wind screaming past them, and they automatically blended RCS with aerodynamic controls as the ship exited or reentered the atmosphere. Not the sort of complex systems a home builder wanted to take on.

So Rutan looked at the most obvious alternative to a winged spaceship: the approach favored by both NASA and the Soviet Union for sending people into space in the 1960s, when Rutan worked at Edwards. This was the capsule design, where the astronaut rode more as a passenger than as a pilot, launched into space on the tip of a missile to reenter without the missile, which simply fell back to Earth as an expensive meteor. The blunt end of the capsule only had to be pointed generally in the right direction as it started its fiery dive into the atmosphere. Its shape would catch the air to keep it oriented blunt end first so that an attached ablative heat shield could take the heat, sloughing off chunks of itself to keep the rest of the capsule cool.

Aside from setting that initial, roughly correct, orientation going into the atmosphere, no input at all was required from the astronaut or was even possible in such a design. Without wings, the craft could be maneuvered only by the RCS operating in space; once it hit the atmosphere, the capsule was just a falling object, bobbing in the supersonic rush of air, but keeping always blunt end down like a cork tossing in a rough sea, until its parachutes popped out. Swaying through the lower atmosphere, blown by the prevailing winds, the capsule headed toward a less-than-pinpoint landing either in a vast expanse of flat wilderness, as with the Soviet designs, or in an ocean, whereupon the armed forces of the capsule's operating government would fan out to look for it.

This didn't strike Rutan as a particularly elegant solution either. As with his homebuilt airplanes, he wanted a craft that could be flown by as many people as possible, as inexpensively as possible. Granted, any spaceship would likely be several orders of magnitude more expensive to build and operate than a fiberglass airplane, but a military-style search and rescue operation after each landing would push the thing into a whole other universe of complication. "I also knew," said Rutan, "that if you put something out there that relied on parachutes to recover . . . you would have a lot of accidents. Parachutes have been around for a long time, and there's been an enormous amount of money spent, an enormous amount of development, and they still have a risk that's not something that you're happy with on something that flies the public."

Still, Rutan couldn't immediately see his way to a better solution, so his first sketches for a Scaled Composites spaceship showed a single-seat capsule on the end of a missile that would be recovered by parachute. His one concession to the X–15 at that time was to make his spaceship air-launched. He'd sling the capsule with attached missile from the belly of a mother ship, like the X–15's B–52, to get up to launch altitude. Using an airplane's jet engines as the first stage of the rocket would let the missile carry far less rocket fuel. And it would be safer. If the rocket engine blew up after the drop, the pilot could eject the capsule from the missile and make an emergency parachute landing.

When Peter Diamandis announced the X PRIZE in 1996, Rutan upgraded his design to accommodate the three people the prize required. And then he had a brainstorm. Since the spaceship would no longer fly just a single pilot, why not make *all three* people on board pilots? Even better, since the capsule design would require little piloting skill, all three could be paying passengers, ordinary people who wanted to experience the wonders of space travel without having to become professional astronauts. Each astronaut-pilot could be trained in ten days—the span of an ordinary vacation. One astronaut-pilot would be in charge of the countdown after the drop from the mother ship, and he or she would hit the button to fire the missile's rocket engine. Once in space, another astronaut-pilot would use a joystick to operate the RCS and orient the craft for reentry. Since this was a capsule, Rutan explained in 1997 to a gathering of home builders at their annual convention in Oshkosh, Wisconsin, "if he messes it up and reenters sideways or backwards he still won't hurt anybody. But he'll have the smoothest ride if he goes straight into the atmosphere." The third astronaut-pilot would deploy the parachutes at the appropriate altitude and communicate with the recovery forces on the way down and after landing in the Pacific Ocean, a hundred miles downrange of Mojave. All of which would give these folks not only the ride of their lives but also the satisfaction of proclaiming themselves true astronauts, pilots of their craft and not just "roller coaster riders," as Rutan put it.

By 1998, Rutan even had a mother ship mostly built. This was a high-altitude research jet plane called *Proteus* he had designed and Scaled Composites had built to carry experiment packages aloft. Since these packages would be carried on the craft's belly, *Proteus* could just as easily drop off its payload—a missile, say— as bring it down again. Without funding to build the spaceship, though, Rutan kept the project on the back burner, recruiting a handful of his engineers at Scaled to work on it as volunteers in their off hours. He also called on one of his former colleagues at Edwards, engineer Bob Hoey, who had worked on the X–15's test flights.

The project fired the imaginations of Scaled engineers Cory Bird and Dan Kreigh, and by 2000, Bird later recalled, "Dan and I were constantly bugging Burt: you know, 'Let's do this thing! Let's build a spaceship!'" But Scaled Composites was in the business of building airplanes, and Bird was convinced that a winged ship was the way to go, not this capsule business. "I kept throwing stuff on his desk that had wings and landing gear on it," Bird said later. He recalled asking Rutan, "'Wouldn't it be nice to be able to launch out of here and land back at the airport?'"

Rutan's space capsule design at the time featured "these feathery looking things," as Bird described them later, "to create a lot of drag." These structures would catch the air and point the craft the right way no matter what direction it was facing when it reentered. It was a very clever idea, Bird allowed, but he didn't see why such devices couldn't be used on something with proper wings and landing gear like any self-respecting aircraft. One of the drawings he gave Rutan showed "some floppy things coming off a normal airplane." Rutan took that one and went off and pondered it for quite some time before Bird heard any more about it. "And all of a sudden," Bird recalled, "he came back in and he was all fired up about some kind of flap that came out the top."

Rutan had figured out that the feather *would* work on an airplane. At least subsonically. He got Bird and a couple of other guys together, had some foam models built, and threw them from the control tower at Mojave Airport. It was like being a kid again, building and flying models. Rutan tried different positions for the top flap, which looked like a speed brake coming up from the middle of the airplane's fuselage. And it did work. But *only* subsonically. When Rutan put the design into his computational fluid dynamics software, the results were horrible in the supersonic region of the craft's flight. And it was the same for every variation he tried. The designs simply didn't create enough drag to allow the craft to make a safe reentry.

Now the problem had a hold on Burt, plaguing him in his waking hours, infecting his dreams. At a restaurant with his fourth wife, Tonya, he'd doodle feather designs on napkins. Sitting in the audience at a charity event, he'd tune out, sketching, fitting the pieces together in his mind, playing the reentry over and over again until at last, on that fateful night, it all just clicked into place.

It was obvious in retrospect. Those hinging tail booms—they

were just like the dethermalizers on the model airplanes Rutan had built and flown as a kid. Radio-controlled models were too expensive for him in those days, and the early radio controls then available didn't work well anyway. So Rutan concentrated on models that could be controlled with wires connected directly to the airplane, and on so-called free-flight models that just took off on their own. A free-flight model had a habit of sailing off, never to be seen again, unless it had an onboard device called a dethermalizer, a timer that would kick the horizontal stabilizers up to a 45-degree angle. At that point, the model would simply stop flying and float to the ground.

From that morning on, Bird said later of Rutan, "It looked to me like he was hooked." Rutan was on a mission now, putting the pieces together for his homebuilt spaceship in earnest. Now he was back in familiar territory. This was an *airplane* ferchrisakes. He knew how to do airplanes. Bird and Kreigh looked at each other and grinned.

It didn't take long for Rutan to rough out the outlines of his two-part homebuilt spaceship. He'd need a bigger mother ship to start with. *Proteus* would have been fine for a one-person capsule-and-missile combo, and *maybe* for a three-place capsule, but for a three-person rocket-powered airplane, he'd need to scale it up. No problem there. That was what he and Scaled Composites *did*. The hinging tail booms and the number of people on board set the basic shape of the spaceship. Now that the ship was a space plane and not just a capsule on the point of a ballistic missile, one of those people would be a true pilot, not just someone punching a button or moving a joystick at the appropriate times, and the other two would just be passengers, along merely for the ride. It would have been nice to enable all three of the astronauts to help pilot the ship, but with this new design, Rutan had the best of both worlds, and there was no going back. He had the control and precision of an airplane coupled with the carefree reentry of a capsule. All he needed now was a financier. And of course a rocket engine.

. . .

Tim Pickens was a child of the first space age. He grew up in Huntsville, Alabama, the "Rocket City," as it was known in the heady days of the first manned space shots. He was born in 1964, the year after the conclusion of Project Mercury, America's first space program, in which fighter jocks rocketed to space in tiny one-man capsules they didn't so much climb into as *wear*, on the top of ballistic missiles originally meant for slinging nuclear bombs halfway around the world. Next came Gemini, two-man capsules that enabled astronauts and engineers to practice for Project Apollo, the moon missions. Gemini featured stays of up to two weeks in orbit, dockings between two spacecraft, and space walks.

Pickens was five years old when Armstrong and Aldrin hopped out of their moon ship onto the lunar surface, and by then he lived, breathed, practically ate rockets. All of those missiles, from Mercury's Redstone and Atlas rockets through Gemini's Titan, up to Apollo's monumental Saturn V, the biggest and most powerful rocket ever assembled, were designed and built at NASA's Marshall Space Flight Center in Huntsville. Pickens's dad was a physicist who worked on the Apollo rockets. Test firings of the Saturn V's F–1 engines, in which the engines were strapped down to a reinforced concrete test stand that was cooled with a deluge of water so it wouldn't burn up in the blast furnace of the engines' exhaust, shook the ground and shattered the air with such awesome power that the windows of the Pickens family home miles away rattled in their frames. Pickens and his oldest brother, Randy, clambered around on discarded rocket parts at the local junkyard after school. Randy built cardboard-and-wood space capsules in their garage, and the brothers sat in them for hours, flipping switches and watching lights left over from project Mercury. They played rockets and they dreamed rockets, and of course it was what they wanted to do when they grew up.

Pickens never got a degree in rocket science; he just lived it right from the beginning, tinkering with rocket parts and rocket-powered vehicles until building and firing rocket motors came as easily as breathing. As an adult he gradually amassed a collection of NASA surplus rocket motor valves and rocket plumbing in his garage, along with the machine tools and electronics and raw metal he needed to build his own motors.

Meanwhile, the Huntsville he had grown up in just sort of seemed to dry up and blow away. "I look around Huntsville and wonder what happened to all the smarts," he marveled. "It's as if we were visited by aliens and they did some neat stuff, then they left without a trace. All that is left is a few rusty test stands and a bunch of lost, dissatisfied engineers." By then, the work of the Mercury, Gemini, and Apollo engineers had been reduced to a museum, the U.S. Space and Rocket Center, where the great rockets stood as monuments, still pointed at the sky but forever cold and silent. Inside the museum building, rocket engines stood on display, all their plumbing and intricate inner workings exposed like exquisite abstract sculptures.

Pickens went to the museum on his lunch breaks, on weekends, on his birthday. He could see his dad's work there, enshrined in the exhibits inside, could sight up the length of the great Atlas rocket that sent the first Americans into orbit, walk down the supine length of the mighty Saturn V. He could run his hands along the cold pipes of the rocket engines, get a feel for how they were machined, welded, bolted. They inspired him. But they also depressed him. These were works of art made by the masters of astronautical engineering, touchstones for anyone who hoped to follow their lead. Much of this hardware could fly today—those very rockets, those very engines. They were so well designed and built that they could be cleaned up, their valves refurbished and resealed, their electronics modernized; they could be hoisted up on pads, fueled, and launched. But they would never fly, and no more of these rockets would ever be built. It was a crying shame was what it was.

One lazy Saturday afternoon in 1998, Pickens and one of his rocket buddies, Glenn May, were sitting around, and, Pickens recalled later, "we just concluded that we needed to take my bicycle and one of my engines—hybrid Plexiglas and laughing gas—I had laying around and put it on there." Eight hours later, they had themselves a rocket bike. It imparted only thirty-five pounds of thrust, just enough to push the rider gently down the road. But that was enough proof for Pickens that the concept was sound. So he built himself a *real* rocket bike, an honest-to-God crotch rocket with a two-hundred-pound-thrust rocket motor capable of blasting him from zero to sixty miles per hour in five seconds flat. That's fast enough to outrun a Porsche in an old-fashioned down-home street race.

Not that even Pickens would try such a thing—that cheapo bike he'd picked up on special at Wal-Mart started to get pretty wobbly past thirty miles per hour, and it hovered just a hair from careening out of control at forty-five. Still, it was Pickens's very own rocket ship. The motor sat right behind Pickens's backside, where a cargo rack would sit on an ordinary bicycle. A button under Pickens's left thumb ignited the rocket motor with a zap of electricity from an onboard battery to an Estes model rocket motor inserted in the rocket's thrust chamber, and he had modified the left brake lever to throttle the thing. For safety he wore a motorcycle helmet and a pair of work boots when he rode it. And, just for fun, a flight suit. "The rocket bike came out of the want to own something that would actually allow me to finally feel what it's like to be accelerated by space propulsion, or a rocket engine," Pickens explained later. "So basically, it's my own subscale space program."

That might have sounded like hyperbole to many people, but Pickens's rocket bike impressed the hell out of Burt Rutan. Here was a guy, thought Rutan, who was so confident in his hardware that he was actually willing to risk his own rear end on it. As Pickens explained later, "You'll find a lot of guys in this business who theorize to death, but will they put their lives on the line? And the answer is 'No.'" Rutan knew that full well by the time he saw Pickens and May's rocket bike in early 2000.

As far as Pickens was concerned, the key to making this miniature rocket ship safe enough to plant your butt right on top of the rocket motor lay in the type of propulsion it used. Most of the NASA and Russian (formerly Soviet) manned rockets used liquid fuel consisting of liquid oxygen ("lox") and either rocket-grade kerosene called RP–1 (Refined Petroleum 1) or liquid hydrogen. Liquid fuels give the most power for a given volume, but their very efficiency can make them volatile. Kerosene in the presence of lox ignites easily, sometimes when it shouldn't, and liquid hydrogen is downright explosive, as the fiery death of the hydrogen-filled *Hindenburg* airship attested in 1937.

Some rockets, like the space shuttle's strap-on solid rocket boosters, run on solid fuel. Though less efficient than liquid-fueled rockets, solid-fueled rockets, which combine fuel and oxidizer in a solid matrix, have the advantage of being fueled during the manufacturing process, making launches easier. They also present their own dangers. As anyone knows who has launched a store-bought model rocket or Fourth of July fireworks, solid-fueled rockets once ignited cannot be shut down. Nor can they be throttled. You just light the fuse, stand back, and hope for the best. Liquid-fueled rockets, by contrast, can be shut down, even split seconds after ignition, if a human operator or a computer determines that all is not going according to plan. They can also be throttled up or down simply by increasing or restricting the flow of fuel and oxidizer through the plumbing that feeds the combustion chamber.

Though less efficient than either liquid or solid rocket motors, hybrid rocket motors combine some of the best features of both. Hybrid rocket motors use a solid fuel and a liquid or gaseous oxidizer, kept apart from each other until the operator opens a valve to allow the oxidizer to flow into the rocket motor—in essence a tube with the solid fuel molded to the inside of it. A small pyrotechnic device then provides the spark, the fuel ignites, and the resulting hot gases shoot through the tube containing fuel and out through the opening at the end that constitutes the rocket's nozzle. As with any rocket engine, the rushing hot gases produce an equal and opposite reaction that pushes the motor (and any attached vehicle) in the other direction. Turn off the flow of oxidizer, and the hybrid rocket engine sputters out immediately.

Pickens throttled his hybrid rocket bike with a valve that increased or decreased the flow of oxidizer—in this case, nitrous oxide from a bottle hung from the bike's crossbar—in response to squeezing or releasing the bike's brake lever. For extra safety, he also had his rocket bike rigged with a kill switch that would instantly shut off the flow of nitrous if his right thumb got jolted loose from a button on his handlebars. For his fuel Pickens chose ordinary roofing tar, purchased from his local roofing supply company. It was safe to handle at any temperature and could be easily and safely stored in his garage.

All in all, the rocket bike made a pretty piece of engineering, Rutan thought when he saw it. It successfully demonstrated all the components of a real spaceship's rocket engine: a motor, an ignition system, an oxidizer delivery system, a throttle, and even some impressive safety features. In fact, this sort of thing might be just what he was looking for for his own spaceship.

Pickens didn't plan it that way; not exactly. But it wasn't quite an accident either.

Rutan saw the rocket bike in Pickens's garage in early 2000, when he was in town to get ideas for his spaceship's rocket engine. He still had no sponsor, but he thought he'd better nail down this crucial piece of his spaceship's design before he got serious about finding one. He hadn't designed propulsion systems for any of his thirty-nine previous aircraft designs. "Some were refit with propellers, some were turbo props, some were turbo fans, some were turbojet," he explained later. "But [for] all of them I would go and buy an engine that was already well-tested, and most cases certified, and I would just do the installation." So

75

it had made sense for him to look for outside expertise to help with the spaceship's engine as well. He had met Pickens in late 1999 after giving a talk at the Huntsville chapter of the Experimental Aircraft Association, and the man had impressed him as someone who knew a thing or two about rockets. After an exchange of e-mails, Rutan had followed his recommendations on rocket shops to visit in Huntsville to explore his options.

Rutan had already narrowed his choices to liquid and hybrid propulsion; solid rockets would be impossible to work with during incremental flight testing—"envelope expansion" as aeronautical engineers called it. "I knew that I couldn't do envelope expansion airplane testing with a solid because it's way too dangerous if you can't shut it down," he said. When you took a new rocket-powered airplane past the speed of sound for the first time, you wanted to be able to shut down that rocket engine in a hurry if anything started to go buggy, not sit tight with your knuckles white on the control stick and hope and pray for the best while the damned thing took its own sweet time burning itself out. "It's way too dangerous if you can't shut it down if you've never flown the airplane to a certain Mach number," said Rutan.

So, on Saturday morning, February 12, 2000, Rutan found himself sitting in a conference room with Chris Barker, head of rocket engine start-up Space America, along with Tim Pickens and several other rocket men then working for the company. Barker popped a tape in the VCR, and they settled back to watch video of a recent Space America liquid rocket engine test. The tape did not inspire confidence, to say the least. Rutan watched as the engine bolted to a test stand, lit up, and then blew itself apart, flinging twisted bits of metal through an inferno of exploding RP–1 and lox. Rutan's eyes, Pickens recalled later, "get bigger than silver dollars." He sat stunned for a moment while Barker's engineers tried not to look too pained. "Whoa!" exclaimed Rutan when he'd recovered his composure. "What happened?"

Barker essayed a nonchalant chuckle. "Well, that was a software problem."

Rutan was not amused. "I just lost my crew!"

"What's that tape doing in there?" asked Barker. He figured they'd just get the right tape in there, and they'd go on with their meeting as if nothing had happened. He didn't realize he'd just blown any chance to contribute to Rutan's new project. The meeting dragged on for another hour and a half, with Rutan picking apart diagrams of the engine, shaking his head at the complexity of the thing, all the valves and plumbing that had to function perfectly in sync to avoid an explosion—a "hard start" as the engineers more prosaically termed it.

One of the engineers pulled Pickens aside as the meeting was breaking up. "I think Burt's really leaning toward hybrids, Tim. It's gonna be up to you to convince him that liquids are the way to go."

Pickens was Rutan's ride back to his motel, and if there was any chance at all for Space America to redeem itself in Rutan's eyes, Pickens was indeed it. But Pickens felt sure there was nothing he could do to change Rutan's mind about the company. So he came clean. "By the way," he told Rutan as they got into Pickens's car. "I'm supposed to tell you liquids are the way to go. But I don't think that's the case for you, for your system. Especially the first time out." Maybe later, on another spaceship that needed more power, Rutan could switch to liquid-fueled rockets, when he and his crew had more rocket experience under their belts. But for now, hybrids would give him the best performance and safety with the minimum of complex parts.

By the time Rutan's next appointment picked him up from his motel that afternoon, Rutan had firmly settled the question in his mind. No way he was going to risk an explosion like Space America's on a piloted craft. His spaceship would run on hybrid propulsion or nothing. Greg Allison, president of the High Altitude Research Corporation, or HARC, showed Rutan his company's test facility on a plot of undeveloped land outside of town and then drove him back to headquarters for the hard sell. Headquarters turned out to be in somebody's garage. Well, nothing wrong with that; as Rutan knew full well, plenty of great projects got built in garages.

But who was this coming out of the house beside the garage? "What are *you* doing here?" Rutan exclaimed when he recognized Tim Pickens.

Pickens had to laugh. "I *live* here."

"This garage is *yours*? But you're with Space America."

"That's a different hat I wear," said Pickens. "This is the HARC hat."

Turns out Pickens was a founder and part owner of HARC, and he moonlighted there when he wasn't working at Space America. HARC was using his dad's land as a test site. And of course this was his garage. Pickens showed Rutan the rocket bike, and it was all over. Poor Allison could only stand by helplessly as Pickens stole his show.

That night Pickens took Rutan to dinner at a fast-food joint called Telini's in a strip mall on Perimeter Parkway and University Drive. There, as the two men bent over their table sketching ideas on napkins, the last major piece of Rutan's spaceship design snapped into place.

First Rutan drew the outlines of his spaceship. Now that he was set on hybrid propulsion, he knew he needed just one tank, for the oxidizer, not the two that would be required for a liquid-fueled rocket. That simplified things. He drew the tank inside the spaceship and stuck the solid-fuel-packed tube that constituted the rocket engine on the back of it. But he didn't know how to secure the tank and motor without a lot of complicated hardware that would unduly stress the spaceship's structure when the engine fired.

Pickens later recalled that he suggested to Rutan an approach he'd been using with his local rocketry club since 1994: build the ship's fuselage around the tank, and then support the motor just by attaching it to the tank. No need for the structural bracing Rutan had sketched on his napkin.

Rutan remembers the conversation differently. The idea was

his, he told me. "My realization that the motor could be supported only by the tank was the breakthrough that excited me that evening since it led to a truly simple, reliable motor configuration." It was actually Pickens, he said, "who felt that I would need to support the motor with struts to the aircraft's structure. I was concerned about that . . . since the aircraft structure is very flexible."

There at Telini's, Rutan calculated the circumference of the nitrous oxide tank, input the planned thrust of the rocket engine (about twelve thousand pounds), and figured from that the total amount of structural stress, or load, the integrated tank and spaceship hull would have to carry. "That's nothing," he said when he had finished the calculations. "I could just RTV this tank in there." Goober it on to the inner surface of the fuselage, in other words, with the same sort of silicone rubber used to weatherize a house.

Regardless of who provided the initial inspiration for the motor assembly, that brainstorming session led directly to *SpaceShipOne*'s final design, and Rutan couldn't have been more delighted. He looked up from his napkins. "Why don't you come work for me and lead my rocket development?" He asked Pickens.

4

SpaceShipOne, Government Zero

SpaceShipOne Wins the X PRIZE

"White Knight with one mobile out to three-zero for departure."

The air traffic controller stared out his windows at the ungainly looking craft, white with red stripes, making its way down the taxiway to runway 30 for the first time. *"What type is it?"* he called back to the airplane outside.

"We're calling it a White Knight,*" replied Scaled Composites pilot Doug Shane.*

"I just gotta ask a question when I see something that isn't . . . uh . . . the same as usual."

Shane chuckled. "What is usual, John?" White Knight *was definitely not usual. One reporter later compared it to a Klingon warship from the* Star Trek *movies. With an eighty-foot wingspan, twin booms with attached tails in the back and landing* gear at the front, and a lozenge-shaped fuselage that rode high above the ground, the ship looked nothing like a B–52 bomber, but as far as Burt Rutan was concerned, it served the same purpose: to carry a rocket-powered airplane to high altitude and drop it to blast out of the atmosphere and into space. Those outthrust booms with the gear on the end, along with a nose festooned with a multitude of round portholes instead of more usual-looking airplane cockpit windows, reminded the ship's builders of a knight in armor. Hence, *White Knight*.

With a whine from its twin General Electric J85 turbojet engines, the *White Knight* turned onto runway 30. Shane pushed the throttles forward, the whine rose to a roar, and the airplane surged forward. Ungainly looking or not, the ship had plenty of power; those engines coupled with its light fiberglass construction gave it a high thrust-to-weight ratio, which meant it could accelerate quickly and climb at a steep angle when it was unencumbered by an attached spaceship.

The *White Knight*'s first flight was on August 1, 2002, just a year and a half after Rutan secured the funding for his home-built spaceship from Microsoft cofounder and billionaire Paul Allen. The craft hit an unexpected snag on its first takeoff when spoilers designed to help the ailerons provide roll control began flipping up and down, requiring Shane to make an emergency landing. But the Scaled crew corrected the problem in short order, and the ship aced its next test flight four days later.

Like so many space enthusiasts, Allen grew up during the first space age of the 1960s, thrilling to the flights of the Gemini and Apollo astronauts. Also like so many others of his generation, he'd assumed he'd grow up into a world where flights to the moon were routine and the first pioneers were already establishing permanent bases on Mars. Now, as an adult sitting on one of the largest fortunes in the world, he'd been looking for a chance to help bring that future about. Funding an X PRIZE–class spaceship seemed like a good start, and Burt Rutan seemed like just the person to pull off building the first private spaceship. Secret meetings over several months sealed the deal in March 2001 for at least $20 million in funding. Since Diamandis hadn't yet come up with the X PRIZE money, Allen and Rutan didn't give winning the prize much thought during their negotiations.

In April 2003, select members of the press received an invitation from Scaled Composites to attend the unveiling of a new aircraft design. The guests included not just reporters but the some of the most influential people in aerospace, including Maxime Faget, designer of the Mercury space capsules; Buzz Aldrin; and Air Force Brigadier General Pete Worden, who had led NASA's DC-X vertical takeoff/vertical landing rocket ship demonstrator in the mid–1990s. The journalists and VIPs watched from folding chairs in the Scaled hangar as Rutan waved his arms in a theatrical flourish and a curtain fell away from the front of the hangar.

A little white ship stood revealed, peering back at the assemblage with the same round cockpit windows as its mother ship. In fact, the pressurized cockpits of the two craft were identical, which meant that every time one of the pilots flew the mother ship, he was also training to fly the spaceship. For that's what it was—a spaceship.

SpaceShipOne, Rutan called it, with the tail number N328KF to denote the lofty ambitions of its designer, who hoped that it would reach 328,000 feet in altitude. It looked a bit like the Bell X–1 in which Chuck Yeager broke the sound barrier for the first time. Bullet-shaped. Small—weighing only three thousand pounds when empty. Its carbon fiber frame featured the same composite construction that Rutan had pioneered for home builders of fiberglass airplanes. Much lighter than ordinary aircraft aluminum or even fiberglass, the ship's airframe was nevertheless stronger than steel. In a design by Dan Kreigh, a spray of stars outlined in a field of blue on the ship's nose slid down along the ship's belly as though it had brought back a swatch of the heavens. The spaceship's white stars on a field of blue, along with the white fuselage and red stripes of the *White Knight*, completed an American flag motif, test pilot Brian Binnie's idea.

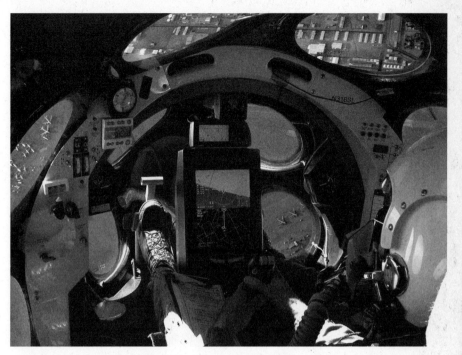

View from the cockpit of the *White Knight*. The layout is identical to that of *SpaceShipOne*. *(Photograph © 2004 Mojave Aerospace Ventures LLC, photograph by Scaled Composites.* SpaceShipOne *is a Paul G. Allen Project.)*

After the unveiling, just to prove that these ships were no mock-ups, Rutan had Mike Melvill and Brian Binnie take the *White Knight* on a joyride to buzz the crowd. The ship dove low, pitched up steeply, and roared skyward with both engines in afterburner. "Dammit!" Rutan laughed. "Those guys are having too much fun! I'm gonna have to stop paying them!"

Still, the hard work had only just begun, and now there was a deadline for winning the X PRIZE. At the suggestion of X PRIZE codirector Bob Weiss, Peter Diamandis had finally gotten the prize money together by taking out a so-called hole-in-one insurance policy, commonly used to ensure prizes for performing some near-impossible feat like shooting a basket from across a basketball court. Iranian-born businesswoman Anousheh Ansari, who had made a fortune in the United States in telecommunications and who desperately wanted a ride to space, had paid the policy's six-figure premiums with her brother-in-law, Amir Ansari. With the Ansaris' backing, Diamandis finally had his title sponsor. The X PRIZE was now the Ansari X PRIZE. According to the terms of the insurance policy, the prize had to be won by the end of 2004, or the insurance company, Bermuda-based XL Capital, would pocket the premiums without having to shell out a dime. It was a risky bet for Diamandis and the Ansaris, and a pretty good one, it seemed, for XL Capital.

SpaceShipOne had yet to fly at the time of its rollout to the press. True to Scaled Composites form, the ship had never seen a wind tunnel; its builders had relied on Rutan's good design sense and computer simulations to trust that it *could* fly. But in contrast to every other Rutan design, there could be no baby steps with *SpaceShipOne*. No taxi tests, no brief hops off the runway, no gradual buildups to altitude. The very first test flight would have to be a drop from *White Knight* at high altitude. That task fell to Mike Melvill, Rutan's most senior pilot. "The most scary flight," Melvill later called that first drop test. And Melvill had spent twenty-five years flying one weird Burt Rutan airplane after another on first flights. "It was a strange-looking little airplane," said Melvill, and "it didn't look like it had enough wing."

When Mike and Sally Melvill moved to Mojave to join the RAF in 1978, Mike was in his mid-thirties and had been a pilot for only a few years. He had come to flying relatively late in life, more or less by accident. He and Sally, both from South Africa, had moved to Indiana to join Sally's two brothers and father in their business machining custom dies for cutting cardboard boxes. Increasing sales required a lot of cross-country travel, so the family decided they could do that most efficiently if one of them got a pilot's license. "My brothers said, 'Well, we're not gonna do it,'" Sally recalled later, "'so, Michael, you're it.'" Mike's career as a pilot got off to a rough start; he suffered debilitating airsickness. But once he overcame that, he jumped in feet first, first purchasing his own airplane and then by getting to work with Sally on building a VariViggen. Building that airplane took three years, and by the end of it, Mike was thoroughly hooked on flying—hooked enough to jump at the chance to work for Rutan in Mojave, enough to sell his stake in Sally's family's business and quit after working there for eleven years. Over the next twenty-five years, Rutan helped hone Melvill into a world-class test pilot.

In all his years of taking Rutan airships on their first flights, Melvill had never been let down by Rutan or any of his designs. Those airplanes had always behaved exactly as Rutan had said they would. Likewise, Rutan knew he could count on Melvill to bring those ships home in one piece, no matter how hairy the ride got. Over the years the two men became close friends, and they trusted each other absolutely. In Melvill's case, he quite literally trusted Rutan with his life every time he took an untested design up for its first flight. But that didn't make him any less nervous about taking *SpaceShipOne* on its maiden flight in the early morning of August 7, 2003. *SpaceShipOne* wasn't any ordinary airplane, and Melvill didn't expect it to fly like one. To begin with, it was a glider, and that's how the Scaled team treated it for those early unpowered flights. All the test pilots brushed up on their glider training, flying sailplanes at a Nevada flight school in preparation for the drop tests.

Drop tests. The idea was terrifying to Sally Melvill. Her husband would ride up to forty-seven thousand feet, a good ten thousand feet higher than a passenger liner, attached to the belly of the *White Knight*. And then, just ... drop. Like a bomb. With no engines on board, Mike would have to fly that thing, if it flew—no one would know for sure until the drop—back to the airport in slow spirals, carefully managing his downward motion, his "energy," in the parlance of pilots, so that he wouldn't come down too fast or stray so far off the mark that he would miss the airport.

Before the ground crew closed the hatch, Sally pinned her lucky horseshoe to the left shoulder of Mike's flight suit. It was a tradition the two had developed over all the years that Mike had test-flown Rutan's airplanes. Mike had had this pin made for Sally during their courtship in their teens back in South Africa, and now she'd taken to pinning it to Mike's flight suit before every dangerous flight. Afterwards she sweated it out on the side of the flight line as Brian Binnie, driving *White Knight*, throttled up the J85s, and the mated ships taxied to runway 30, turned about, and then roared off down the runway and lifted into the brightening sky.

It took an hour for the *White Knight* and the attached *SpaceShipOne* to spiral up to the rarefied air at forty-seven thousand feet. At this height, they were clear of most of the atmosphere, and that would make it far easier for the rocket motor that would be installed in the spaceship to boost the ship the rest of the way to space. Aside from the agonizingly long wait for his drop, the practical upshot for Melvill was that he got cold. Very cold. The temperature outside the spaceship's thin carbon fiber walls dropped to minus 60 degrees Fahrenheit or lower. The ship had only batteries for power, and there wasn't enough to run a heater. Some of the *White Knight*'s engine exhaust heat got funneled to the back of the spaceship, but there in the nose of the craft, his feet stuck way out in front on the rudder pedals, Melvill shivered while he waited.

Finally came the word from Doug Shane back in mission control on the ground: the two ships were go for launch. Melvill flipped a switch on his control panel to unlock the latches holding his ship to the mother ship. Brian Binnie in the cockpit of the *White Knight* above him flipped an identical switch. After a brief countdown, Cory Bird, riding in the *White Knight* behind Binnie, pulled a handle. The latches opened. Suddenly lightened, the *White Knight* lurched upward. In *SpaceShipOne*, Melvill lifted from his seat, weightless as he and his ship went into free fall.

The feeling of falling lasted only a moment, however. The ship's wings grabbed at the thin air and hung on. Melvill toggled a switch on the top of the control stick to move the horizontal

The *White Knight* carries *SpaceShipOne* on a captive carry test. *(Photograph © 2004 Mojave Aerospace Ventures LLC, photograph by Scaled Composites.* SpaceShipOne *is a Paul G. Allen Project.)*

stabilizers, the trim stabs on the airplane's twin tails, to level the airplane out, and then he was flying. Silently, without an engine, but flying nevertheless. The crazy little airplane actually flew. "Flies like a dream," in fact, Melvill told the ground.

On the tarmac back at the airport, Rutan heard that on a handheld radio. He grinned up at the sky, watching his bird wing its way home. "Damn, that's a good airplane!"

Melvill touched down in a perfect landing and was greeted with a tearful hug and kisses from Sally. The world's first private manned spaceship was now a flying reality. But it still needed a rocket engine.

Tim Pickens had gone to work at Scaled Composites in April 2001, just after Paul Allen had committed to funding the *Space-ShipOne* project. His first task had been to make a presentation on hybrid rockets to Allen, complete with a demonstration firing of one of his portable "suitcase" hybrid rocket motors. Now ensconced in the Mojave Desert with the rest of the Scaled crew, he was a long way from his home stomping grounds in Huntsville and from his wife, Melanie, and daughter, Sarah, who had stayed behind. As always, Melanie supported Tim in his rocket flights of fancy. But she was now fifteen years into a career as a schoolteacher, and she knew better than to jeopardize that for yet another rocket project whose financial promise was speculative at best. Pickens quickly enough, like everyone else who went to work for Burt Rutan, found out that there was absolutely nothing to do in Mojave *but* work. He threw himself into work on *SpaceShipOne*'s hybrid rocket motor wholeheartedly and tinkered with his rocket bike, version two, on the side.

Pickens and Rutan decided early on that even though the propulsion project would be managed at Scaled, and that Scaled would build the oxidizer tank and the rocket motor's "throat, case, and nozzle" (the CTN, they called it), they'd outsource the rest. Contractors would do the rocket science—load the case with synthetic rubber rocket fuel and build the ignition system, the oxidizer injector, and the filling and dumping system for the nitrous oxide tank. For a while, Pickens held out hope that the company he'd founded in his garage, HARC, would get the contract, and he'd be able to go home *and* work on *SpaceShipOne*. But it was not to be.

Rutan and his managers selected two companies to compete for the rocket science—SpaceDev of Poway, California, and Environmental Aerospace Corporation (eAc), of Miami, Florida. By then, nine months after Pickens had gone to work for Scaled, the heavy design work was complete, and "and now it was herding these cats," as he said later. Managing contractors and the day-to-day grind of getting the engine operational wasn't the kind of work that fed Pickens's soul. Sarah was growing up without him back in Huntsville, and, he told Rutan, "I really miss my wife."

"Is that all?" Rutan replied with a grin. "We can get you one of those!"

Pickens went home. He continued with the project as a consultant, staying in touch by e-mail and flying out for engine tests, but his major role was finished.

Meanwhile, eAc got the contract for the nitrous filling and dumping system, but the competition between it and SpaceDev for the rest of the work got off to a shaky start. The motor SpaceDev had loaded with fuel suffered a valve failure and refused to stop burning after its first run on Scaled's desert test stand. An airport fire truck had to douse the flames before they consumed the CTN. The eAc-fueled motor had a good first firing but refused to light at all on subsequent tests. The two companies eventually smoothed out the kinks, and both got good test firings. Good enough to give Rutan the confidence that neither engine was going kill one of his pilots. "And as it turns out, both of those companies came up with components that did work," Rutan said later. "I ended up going with the one that had the lower bid, and it happened to be the one with a couple or three percent [better] efficiency. They were very close." SpaceDev won the competition.

Six more glide flights and final testing of the rocket engine later, the spaceship was ready for its first powered flight. On December 17, 2003, one hundred years to the day after the Wright Brothers flew the first powered, heavier-than-air craft, Brian Binnie climbed into *SpaceShipOne* and rode beneath the *White Knight* through that hour-long spiral to launch altitude, just as he had for a glide flight a couple of weeks before. The outcome of this flight, however, would be quite different from the previous one.

After the drop and the free-fall plunge, Binnie flipped open a couple of switch guards on his left armrest and in rapid succession hit the engine "arm" and then "fire" switches.

Boom! The fifteen-thousand-pound-thrust rocket engine fired and slammed Binnie back into his seat with the force of three and a half gs. The horizontal stabilizers were set for an aggressive climb—for the rapid pull-up that would be needed to reach space on a full-up flight. *SpaceShipOne* shot upward, adding another g to Binnie's chest. Binnie hung on to the stick, nudging the trim control at the top to keep the nose pointed just slightly down from the vertical. At the same time he worked the rudder pedals to keep the ship from rolling. He dared not look out those little round cockpit windows; he was moving too fast to allow himself to be distracted from the display in front of him.

The display had been designed by Scaled pilot-engineer Pete Siebold. Called the Tier One Navigation Unit, or TONU, it showed *SpaceShipOne*'s trajectory superimposed over a red line showing the ship's ideal flight path. Binnie's job was to stay on that path.

He rode the bull, as he described it later, with the rocket's engine howling in his ears, the ship bucking and shuddering. By the time Doug Shane back in mission control had finished counting out fifteen seconds, *SpaceShipOne* was traveling faster than any private craft ever built. Faster than the speed of sound, in fact—faster than Chuck Yeager booming over Edwards Air Force Base in the desert just to the south fifty-six years before.

Then a timer switched the engine off, and Binnie was thrown forward with 2 g's as the thin air around the ship resisted its forward motion. The sky had darkened to a deep mauve around him. The Earth bowed slightly at the horizon as though seen through a wide-angle lens.

At 67,800 feet, Binnie reached the top of his climb, arced over the top, and started down. With the tail booms hinged upward and locked into place, the ship went into feathered mode, falling flat on its belly, bleeding off speed, until Binnie swung the tails back down at 33,000 feet, airliner height, and turned *Space-ShipOne* into the glider he had flown earlier that month.

Piece of cake, Binnie thought. The hard part was behind him, that pulse-racing rocket motor firing and the accompanying noise and buffeting, and God knew what gremlins lurking. And he'd just paved the way for *SpaceShipOne*'s first spaceflight.

But no flight is over until the wheels come to a stop at the end of the runway.

Mike Melvill, flying chase with pilot Chuck Coleman in Coleman's Extra 300 prop plane, called out Binnie's altitude to him as he came in for a landing. The configuration of *SpaceShipOne*'s round porthole-style windows, designed to reduce stress on the pressurized cockpit, hindered the pilot's ability to judge his height as he landed. He found the help of a chase plane most welcome.

Binnie felt a wing drop on him. He tugged the stick to bring it back up, and then the other wing dropped, and then the airplane began to wobble back and forth as he fought with the stick.

The airplane wasn't behaving the way it had in the simulator and on his previous glide flight. He thought it was about to stall, so he put the nose down to get more speed, and then the runway came up, far too fast. He pulled back the stick to bring the nose up for his landing flare, but it was too late, and with no engine, there was no chance for a go-around

Bam! He hit hard. The left landing strut snapped, and the ship slewed sideways, slid off the runway, and came to rest in a rolling cloud of fine red desert sand.

Unharmed in the cockpit, Binnie cried out in frustration. "Damn it!" He punched the air. "Damn it!"

That was it, he thought. He'd blown it. He was certain he'd have no second chance to fly the spaceship. His career as an astronaut was done before it had started, and the thrill of that wild rocket ride only minutes before faded like the blue sky over Mojave at sixty-seven thousand feet.

It was a reasonable assumption; time was too short before the X PRIZE expired. One more slip up like that could cost Scaled Composites and its benefactor, Paul Allen, $10 million. Whether or not the hard landing was Binnie's fault, it might just be prudent to let someone who *hadn't* cracked up on landing fly the prize flights.

Pete Siebold got the next powered flight the following April.

A young computer expert as well as a pilot, he'd come to work for Scaled in the late 1990s, seeing a perfect opportunity to nurture his love of both computers and flying machines. His two passions found a perfect blend in his work on the TONU and the *SpaceShipOne* flight simulator, which he also designed and helped build.

For Siebold's flight, *SpaceShipOne* carried its full load of nitrous oxide for the first time, making it heavier than ever before. The extra weight made it stall at a higher speed. After the *White Knight* dropped him off, Siebold tried to slow down to the ideal speed to light the rocket, but the spaceship teetered dangerously close to losing lift. He debated what to do with mission control. Was the ship revealing a gremlin that would get worse when the rocket engine fired? They agreed that Siebold should light the rocket motor anyway, rather than risk coming down hard with a heavy load of fuel on board, and Siebold finally hit the "fire" switch. But in the time it had taken to discuss the problem, Siebold had dropped too low for the rocket motor to be able to send him to the extreme altitude that would allow him to test the ship's ability to reenter at supersonic speeds while in feathered mode, as planned for this flight.

That task was left to Mike Melvill on the next powered flight in May. Nothing was ever routine about a ride on *SpaceShipOne*. This time the TONU flickered during the boost, then cut out completely, leaving Melvill flying blind with the engine still howling at his back.

For an instant he considered shutting the engine down and aborting the mission, once again leaving *SpaceShipOne* short of its planned altitude. Instead, he took his hand off the engine switches and, looking out the windows, used the horizon as his guide for keeping the ship on course nearly straight up. He was rewarded with a view of the sky above him turning from deep blue to black.

After peaking at 211,400 feet, Melvill made a feathered reentry, brought the ship out of feather mode at 55,000 feet with no problem, and touched down in a perfect landing. Nowhere to go now but up.

June 21, 2004. Just before dawn, Mike Melvill climbed into *SpaceShipOne*. This time the ship had a full load of rubber fuel packed into its rocket engine and plenty of nitrous oxide in its tank to finally let it live up to its name. This time, if all went well, Melvill and *SpaceShipOne* were going to space.

During the night, something like twenty thousand people had streamed onto the airport grounds in cars and RVs. It was the first time in thirty-two years of aircraft development that Rutan had ever invited the public or press to watch a test flight. One group of space enthusiasts had imported a trio of DJs from Los Angeles, along with strobe lights and enough techno music to rock through the night. Some, like a woman in tight-fitting silver spandex and antennae on her head, wore spacey costumes. The engineers and managers of XCOR Aerospace, a small company building its own spaceship just down the flight line from Scaled, had opened their hangar doors and thrown a party of their own. There was a celebration going on, a regular space Woodstock.

Just after 4 a.m., Burt Rutan slipped away from the launch preparations to help direct traffic and see the crowds for himself. He reveled in the outpouring of support for his years-long dream, now about to reach fruition.

Melvill saw the crowds when he taxied out beneath the *White Knight*, and, he said afterwards, "I was absolutely in a state of shock at the amount of people standing along the fences screaming and yelling and waving. But it was great."

When they reached forty-seven thousand feet, Melvill and *White Knight* pilot Brian Binnie went through their prelaunch checklist, and then, after a quick countdown, Matt Stinemetze, acting as flight engineer in the *White Knight*, pulled the lever to drop Melvill. As he fell Melvill unguarded the arm and fire switches and lit the rocket motor. After a second's delay while oxidizer flowed to the engine and the igniter touched off the fuel, Melvill was slammed back into his seat with the by-now-familiar 3 g's of force.

This time, the ship rolled 90 degrees to the left. That wasn't supposed to happen. Melvill stomped on the rudder pedal to roll the ship back to the right. At the same time he reached for the switch to shut down the engine, afraid he was about to lose control. He held on, though, and the ride smoothed out until the engine made the expected transition from liquid to gaseous nitrous oxide. The ship shuddered and bucked, oscillating between full and partial thrust until the transition was complete.

After the motor cut off and he sailed out of the atmosphere, Melvill tried to move the trim stabs from their nose-up setting to their more centered landing position. But only one of them rotated on its electric motor. The other refused to budge, remaining stuck in its nose-up position.

Back in mission control, Scaled aerodynamicist Jim Tighe frowned at the telemetry data on his laptop. "This is not good," he said to Doug Shane over the mission control communications loop.

Melvill couldn't hear Tighe's comment to Shane. All he heard was Shane telling him to switch to the backup power system for the trim stabs and try again to align them, but he didn't have to be told that he had a real problem. "I had a situation," he recalled later, "where if I couldn't have fixed it, I couldn't have survived the reentry because the two stabs were just way out of alignment with each other." It didn't help his state of mind that he had heard a series of loud bangs from the back of the ship shortly after the engine burned out—caused by harmless chunks of unburned fuel flying out of the rocket motor, as it turned out. At the time, though, Melvill was half convinced that a sizable piece of the back end of the ship had blown off, and that's why the stabs weren't responding. But he kept his cool. "I'm trying," he told Shane simply, without giving voice to his anxiety.

The motor had shut down when it burned through its fuel

Scaled Composites mission control during *SpaceShipOne*'s first spaceflight on June 21, 2004. Burt Rutan sits in the foreground, with Doug Shane to his left. Engineer Clint Nicols monitors life-support systems beside Shane. Behind them (left to right) are forn X–15 engineer Bob Hoey and Scaled aerodynamicist Jim Tighe. *(Photograph © 2004 Mc Aerospace Ventures LLC, photograph by Scaled Composites.* SpaceShipOne *is a Paul G. Allen F*

at 180,000 feet—Melvill needed all the power he could get if he was to reach space, so he had let it shut itself down. Because of his altitude-stealing roll at the beginning of the rocket boost, he only just barely made it. The spaceship coasted to an apogee of 328,491 feet, clearing the 100 kilometers to space by just under 500 feet.

After a few seconds, the trim stab mysteriously started moving again. Melvill hastily centered the two stabs, and he agreed with mission control that he wouldn't touch the trim settings again until he was safely on the ground and they all figured out what the problem was.

The problem turned out to be a simple safety feature on the trim stab's electric motor, designed to shut the motor down to protect it from damage if it was overworked. The safety feature had kicked in when Melvill had run the stab against the limit of its range. As it was designed to do, the motor began working again after a pause of five seconds or so.

At the top of his flight, Melvill finally had a little time for sightseeing. He saw the California coast spread out below him, all the way south to San Diego and north to Bishop. Brilliant white cirrus clouds sweeping inland over the mountains looked exactly like a thick cover of snow glinting in the raw sunlight, unfiltered by the atmosphere. The colors and textures of the ocean, the clouds, the mountains, and the desert below dazzled him.

While he was still weightless, Melvill reached into a pocket of his flight suit and fumbled out a clutch of M&M's. The candies tumbled through the air in the cabin, spinning and bouncing without falling. Video of that sight, captured by a cockpit camera and played over and over again later that day at the press center at the airport and in the XCOR hangar, kept us enthralled on the ground.

After a perfect landing, made despite the fact that his wild ride had sent him twenty miles off course and disabled the trim stabs, Melvill jumped from the spaceship and pumped his fists at the blue desert sky in triumph. He was America's newest astronaut, and the first to reach space without government help of any kind. The occasion had particular significance, since NASA's space shuttles had been grounded for more than a year because of the *Columbia* accident. *SpaceShipOne* was the only manned American craft to leave the atmosphere in 2004.

After taking some questions from the press stand that had been set up along the flight line, Melvill climbed atop *SpaceShipOne* and let himself be towed farther so he could wave to the cheering space fans lining the fences. One of those fans held up a sign, and Rutan jumped from the pickup truck towing the spaceship to retrieve it and give it to Melvill. Melvill held it up over his head, grinning broadly. "*SpaceShipOne*," the sign read. "Government Zero."

Freshly minted astronaut Mike Melvill hugs *SpaceShipOne* financier Paul Allen. *(Photo: Wendy Kagan)*

The way cleared for the X PRIZE flights, the Scaled team got to work beefing up *SpaceShipOne*'s rocket motor and trimming weight off the ship so it could carry the weight of not only a pilot to space, but enough ballast to stand for two passengers besides, as required by the X PRIZE rules. The spaceship pilots wouldn't know who would get the first X PRIZE flight, or X1 as it was known by the X PRIZE organization, until just three weeks in advance, so all three went into training for it.

Only a month before the scheduled launch on September 29, 2004, Pete Siebold faced two life-changing events: the birth of his second child, and a cancer scare. Within a couple of weeks, an oncologist determined that his enlarged spleen was in fact benign, that he didn't have cancer. But by then he'd been thrown off the horse he'd been riding, as he put it later.

With only two weeks to go before the prize flight, Siebold, now the X1's designated pilot, had some hard thinking to do.

He'd already dodged one bullet he had no control over: the threat of a fatal disease. Was it really fair to his family and his new-born son to deliberately step in front of another bullet so soon, or rather ride one in the form of a barely tested temperamental spaceship? Then, too, an enlarged spleen carried a risk of rupturing under high acceleration. It was a tough choice, but he finally concluded that it "would be unprofessional on my part to go fly knowing I had something that could possibly jeopardize the flight."

Mike Melvill got the flight assignment at the last minute, and once again Sally Melvill had to bid her husband a tearful good-bye, pin her lucky horseshoe to his flight suit, and watch as the hatch of a Burt Rutan special closed between them.

Once again, thousands of spectators gathered before dawn on the airport grounds along with hundreds of reporters and photographers, broadcasters with their satellite trucks, and thousands of volunteers to keep chaos at bay. Once again we watched as the *White Knight*, flown by Brian Binnie, took off with *SpaceShipOne* slung beneath with Mike Melvill along for the ride to forty-seven thousand feet. It was all beginning to seem almost routine.

That is, until near the end of Melvill's rocket boost, when *SpaceShipOne* suddenly began twirling about its axis like some rocketized whirligig. As before, Melvill kept the rocket burning, determined to make the X PRIZE altitude, at the same time fighting to control the ship with all his skill as a test pilot. The air was too thin for the ship's aerodynamic surfaces to do more than slow the spin, so Melvill left the atmosphere still rolling.

Out of the atmosphere, Melvill blew through most of both air bottles feeding the RCS, which consisted of little compressed-air thrusters designed by Tim Pickens. That did it. When the roll finally stopped, Melvill unlimbered a digital camera and began snapping pictures out the windows.

Once again, Melvill made a flawless landing, this time deliberately twirling the ship one final time in a victory roll before touching down.

One X PRIZE flight down, one to go—and it had to be accomplished within two weeks for *SpaceShipOne* to bring home $10 million. So far, the spaceship had not managed a powered flight without a problem. Private manned spaceflight, it seemed, was still far from routine.

Brian Binnie had been tormenting himself for months over his crash landing on *SpaceShipOne*'s first powered flight back in December. He felt awful about it, even though he'd flown the rest of that extremely demanding flight perfectly and his hard landing had resulted in nothing worse than minor damage to a landing strut. It was also natural for him, an exacting test pilot, to take responsibility for the mishap despite aerodynamicist Jim Tighe's assurances that the fault really lay with modifications made to the spaceship just before the flight.

The Scaled engineers had installed devices on the ship's control surfaces called viscous dampers—small hydraulic cylinders—in an effort to head off a possible flutter problem. Trouble was, the dampers caused a problem of their own: an unforseen tendency to bind up the controls. "Brian would have had to apply about twelve pounds of force on the stick to break it free," Tighe said later. "Which isn't something you want to do when you're trying to do fine maneuvering."

Not only had the ship never flown with the dampers, but there was also no way to duplicate their effects in the simulator—which made landing with them an unexpectedly serious challenge. "Brian did a phenomenal job considering the circumstances," Tighe said. To correct the problem before the next flight, the engineers reduced the dampers to a third of their initial power and wrapped them in electrically heated blankets to keep the oil in them properly fluid after exposure to the freezing temperatures of the upper atmosphere.

Brian Binnie fell in love with airplanes while growing up in Aberdeen, Scotland. He was born in West Lafayette, Indiana, in

1953 while his father was a visiting physics professor at Purdue, but the family moved back home to Scotland when Binnie was five years old. When Binnie was eight, his father got him a balsa wood glider, and Binnie spent hours in the backyard launching the glider and dreaming of one day flying a real airplane. When the 1960s came around, he caught the space bug too. Of course, at the time, no Scot had a prayer of becoming an astronaut. But Binnie was also an American, leaving him a glimmer of a hope that he could one day reach for the stars. That hope got stronger when he was fourteen and the family moved back to the United States, this time to Boston.

Binnie had a rough time in high school, where his Scottish accent made him an outsider and hence the butt of jokes by some of his classmates. So he made a commitment that was to serve him well in the years to come: he would push himself to do whatever it took to achieve his goals. In this case, he worked hard to erase his Scottish accent, and today not a trace of it remains in his carefully measured speech. He also developed an abiding love for America and the opportunities it presented him—opportunities he could never have realized in Scotland.

After earning undergraduate and graduate degrees in aerospace engineering, Binnie signed on for what became a twenty-year career as a naval aviator. During that time, he flew more than thirty combat missions in the Gulf War and mastered the most demanding of all jobs in aviation, that of test pilot. It was the way many NASA astronauts had blazed a career path toward space. And although that was not to be for Binnie, his job nevertheless gave him all the satisfaction he needed—until 1998, when his superiors wanted to retire him to a desk at the Pentagon. Binnie just couldn't stomach the idea. Nor was he inspired by the second career choice of many of his colleagues: airline captain. After he'd spent two decades flying some of the world's most advanced airplanes, nothing seemed duller to Binnie than the prospect of flying a desk or a glorified bus.

Fortunately, one of Binnie's friends in the Navy, fellow test pilot Marti Sarigul-Klijn, held out a third possibility. It was an extreme long shot a best, but it looked like Binnie's last chance to get to space. A start-up called Rotary Rocket in Mojave, California, had begun work on what its engineers hoped would be the world's first privately built spaceship. Sarigul-Klijn had left the Navy to sign on as the venture's test pilot, and the company needed one more. Binnie didn't have to think twice. With his Navy flying career finished and the possibility of getting to space as a NASA astronaut never more than vanishingly small, he was ready to roll the dice, even if it meant moving his wife and three kids to the middle of nowhere in the Mojave Desert. Besides, that airline job would still be there if the gamble didn't pay off. Binnie's wife, "Bub," cheerfully accepted the relocation, just as she had the many other moves Binnie's military assignments had required.

Rotary Rocket went bust after only a couple of years. By then, however, Binnie had become golf buddies with Burt Rutan, whose company built the airframe for Rotary's test vehicle. When Rotary folded, Binnie simply walked up the flight line and into a job as test pilot for Scaled. And his chances for reaching space suddenly got a whole lot better.

That is, until he crashed the spaceship.

He thought it just barely possible that Rutan might give him a second chance, but he couldn't bring himself to ask because he knew he might not like the answer. So he spent nine and a half months on edge, continuing to train in the *SpaceShipOne* simulator for a flight he didn't know whether he would ever get to make. He took up running, punishing his body to keep his mind off the question he didn't dare ask, even though his bad knees made it agony and he hated running even more than he hated flying a desk. The discipline kept him focused. Having a front-row seat for both of Melvill's spaceflights, however, didn't help him feel any better.

As Melvill saw it, "Brian had sort of fallen out of favor after his landing accident." He thought Binnie needed to prove himself to regain his standing, and he thought he knew just the way for him to do it. As it turned out, Melvill's Long-EZ made an excellent *SpaceShipOne* simulator. All the pilots had to do was fly it into final approach at 145 knots indicated air speed with the engine idling and the speed brake and the gear down, and it would descend at exactly the same rate as the spaceship. To complete the simulation, they made a cardboard mask to put over the Long-EZ's window and duplicate the spaceship's limited view through its portholes. Over the course of a month, Binnie flew eight-four approaches and landings in the Long-EZ. By the end of it, Melvill felt comfortable telling Doug Shane, in charge of *SpaceShipOne* pilot selection, "Brian has worked really hard on this, and he's right up to speed. . . . He deserves another shot." Shane agreed. But Rutan had the ultimate authority, and still he hesitated—right up until after Melvill's X1 flight.

The final X PRIZE flight, X2, carried the highest stakes of any *SpaceShipOne* flight. Far more rode on it than the $10 million prize. Rutan wanted to start the first commercial spaceline, and the means with which to do it lay almost within his grasp. Richard Branson waited with pen poised over checkbook to give him the capital to build a fleet of passenger spaceships for his newly formed Virgin Galactic. But wealthy space tourists weren't going to fork over $190,000 to ride a twitchy experimental spaceship with an unfortunate propensity for flying off course, spinning while traveling faster than a rifle bullet, or cracking up on landing. To seal the deal on a new space age and win development money from Branson on top of the $10 million X PRIZE, *SpaceShipOne* had to demonstrate something it had never done before: that it was capable of a perfect spaceflight, one that would give passengers the thrill of a lifetime without scaring the hell out of them.

As an experimental ship, *SpaceShipOne* had the job of providing the flight data that would make *SpaceShipTwo* safer and, to passengers, reassuringly predictable. But Rutan couldn't wait for *SpaceShipTwo*. He had to deliver the goods *now*. He couldn't even spare the time to further modify *SpaceShipOne*. He had to make that perfect flight as soon as possible after X1, while the world's media were still watching. All the ship's idiosyncrasies and imperfections had to be corrected not in a redesign of the ship—that would come with *SpaceShipTwo*—but in the way the pilot flew it. The *pilot* had to make that perfect flight. More than that, he had to make it look easy.

It took all of a day and a half for Jim Tighe and Pete Siebold to analyze the data from X1 and determine that, yes, it was just possible for a steady hand and split-second timing to pull off that elusive perfect flight. The day after Melvill's flight, Binnie got his second chance. Sitting in the meeting at 4:30 Thursday afternoon, September 30, when the decision was announced, Binnie felt as though he'd been "found not guilty in a courtroom," as he later put it. He didn't have time to dwell much on his good fortune, though; he had work to do.

On Friday morning Binnie climbed into the simulator while Doug Shane manned his usual post as the voice of mission control. Shane's job was to watch his displays, identical with those in the simulator and the spaceship, and talk Binnie through the delicate maneuvers that would let him win the X PRIZE without a glitch.

Over and over throughout that day and through the weekend, Binnie and Shane practiced flying the ninety seconds between the drop from the *White Knight* and punching out of the atmosphere, concentrating on the critical pull-up maneuver—"turning the corner," the pilots called it. They pulled twelve- to fourteen-hour shifts while the rest of the Scaled crew worked around them, installing a new rocket motor in the spaceship and loading it with the X PRIZE ballast.

Binnie didn't sleep much the night before the flight, and as early as his preflight briefing was, at 4:30 a.m., it couldn't have come soon enough for him. He got to the Scaled hangar at 4:00.

At dawn on Monday, October 4, 2004, the *White Knight*, with Mike Melvill at the controls and *SpaceShipOne* project engineer Matt Steinmetz in the back seat and the spaceship hooked to its belly with Binnie inside, taxied down the flight line. Once again Binnie was awed by the number of people who had turned out to see *SpaceShipOne* scratch the sky. But this was the first time he'd seen the crowds from the cockpit of *SpaceShipOne*. "I hope I have a good day," he thought, "because there are an awful lot of people there!"

Just as the red light of dawn began to spill over the Tehachapi Mountains, the two ships turned at the end of the flight line to face back the way they had come. Melvill throttled up the *White Knight*'s engines, and the two ships picked up speed until they roared past the crowds and leaped into the sky. Then Binnie settled in for the long wait while they circled to launch altitude.

For the *SpaceShipOne* pilots, that hour-long climb was by far the most anxiety-ridden phase of the flight. There was absolutely nothing for the pilot to do but review all the things that could go wrong after the *White Knight* dropped him off. Anxiety began to gnaw on the edges of the mind, shading into fear. Compounding the pilot's discomfort was the deepening chill that crept into the cockpit as the mated ships climbed. "Frosty toes were a trademark of SS1 flights," Binnie later wrote. "Numb toes and a numb mind." For once, though, Binnie welcomed the anxiety and the sight of frost forming in the nose of the ship; both were proof of his second chance.

When at last the wait was over, Binnie clasped his hands in front of him, closed his eyes, and breathed a silent prayer.

Melvill flipped the switch in *White Knight*'s cockpit that unlocked the hooks securing the spaceship. A light glowed amber on his instrument panel. "Okay, I got one yellow," he radioed.

Binnie toggled an identical switch on his panel.

"And I've got two yellows," called Melvill. "And stand by with thirty seconds. . . . Twenty seconds to go. . . . Ten seconds. You ready, Matt? Three, two, one, release."

Stinemetze pulled the release lever, and *SpaceShipOne* fell free of the *White Knight*.

Binnie hit the switches to arm and then fire the rocket motor in quick succession. "I didn't waste any time," he later recalled. "It was release, arm, fire, *boom*."

In the *White Knight* Melvill and Stinemetze were still close enough to actually hear the rocket motor fire, something the *White Knight* crews hadn't been able to do on previous missions. Melvill banked the *White Knight* hard right as *SpaceShipOne* screamed past. "Holy crap, that was close!" blurted Stinemetze over the intercom.

In the simulator, the spaceship pilots couldn't hit the *White Knight* on ignition if they tried, so they had stopped worrying about that possibility as they raced to light the rocket motor closer and closer to drop time—the faster they could fire the motor, the higher they would go before shutdown. "You drop off with the trims set for a pretty aggressive turn," Binnie explained later. "Getting the nose up quickly means all that rocket thrust is giving you altitude as opposed to just accelerating you out horizontally. That's goodness."

Not so good would be for the rapidly accelerating and climbing spaceship to catch up to the *White Knight* and hit it like a missile. Analyzing the flight afterwards, Binnie discovered that, as he put it in his typically understated way, "if the *White Knight* had continued to fly straight ahead for whatever reason, it looks like the two vehicles likely would have touched each other. And that would probably be the end of things." That's a nice way of saying that if Melvill hadn't peeled off so quickly, there would have been a fiery midair collision that would have killed all three people on board and closed the era of commercial spaceflight even as it began.

Melvill didn't see it that way. "No, no, there was no close call there," he insisted later. "There was plenty of separation." Besides, "There's no way in the world that you could do it such that you could hit the *White Knight*. It'd be physically impossible because you're level and you out-accelerate the *White Knight* by such an enormous amount that even if you turned instantly, you'd still go out in front of it and not through it." In fact, said

Melvill, "I thought it was pretty cool" to have such a close view of the spaceship taking off.

One of the *White Knight*'s J85s did quit right after the drop. No big deal—the airplane was at the upper limit of the engines' altitude range, and they occasionally sputtered out on these flights. So Melvill decided to have a little fun at Stinemetze's expense. "I freaked out old Matt by asking him for the engine relight checklist." There was no such thing because the engines had no starter motors. All the *White Knight* pilot had to do to restart one was to make sure rushing air kept it spinning as he descended to a lower altitude.

Stinemetze, already shaken by the apparent near collision, pawed through his papers. "I don't seem to have anything like that," he said, trying to stay calm.

Melvill laughed. "I didn't think you would."

"You son of a bitch!"

Meanwhile, in the cockpit of *SpaceShipOne*, Binnie was entirely focused on the task he'd spent the past three days training for in the simulator. The trick to avoid spinning the ship, as Melvill had done on his last flight, relied on precise movements of *SpaceShipOne*'s trim stabilizers, or "stabs," on its twin tail booms.

The pilot controlled these big, electrically powered control surfaces through a thumb switch on his control stick. Thumbing the switch toward him lowered the leading edges of the stabs, giving the ship "nose-up trim." Rocking the switch forward raised the stabs' leading edges, giving the ship "nose-down trim."

Normally, trim controls on an airplane merely keep it stabilized in flight, while smaller control surfaces on the tail, called elevators, control most of the up-and-down pitch of the plane's nose to help it climb or descend. The pilot operates the elevators by pushing forward on the stick to nose the plane down, or pulling back to bring the nose up. But *SpaceShipOne*'s elevons, serving the same function and operated through a system of mechanical push rods, became less effective at supersonic speeds. After the rocket boost, aerodynamic forces working on the control stick made it difficult to move, and only the motorized trim stabs could easily shove the plane around.

"Good light," Binnie gasped out as *SpaceShipOne*'s rocket motor slammed him back into his seat with the force of 3 g's. With the stabs set for a rapid climb, the ship pitched up as it accelerated, hitting the speed of sound in about ten seconds. The craft shook and vibrated as it crossed the sound barrier, and then Doug Shane's calm voice back in mission control reminded Binnie of his next task. "Okay, start the nose-down trim."

This step was critical to keeping the ship from rolling. Binnie had to slow the rate at which the ship pitched upward toward the vertical, or he'd lose control as the rocket burned through its fuel and began to apply thrust to the ship slightly off center. Binnie pushed the switch under his thumb forward. His goal was to spend the next forty-five seconds into the burn "milking the nose," as he later put, to gingerly bring it up from a 60-degree pitch angle to 80 degrees.

"Looking great at twenty seconds," called Shane. "Okay?"

"Doing all right," said Binnie. "Little lateral oscillations now."

"I see that. Thirty seconds. A little nose-up trim's probably okay now."

By the end of that first minute of the flight, the ship was traveling at three times the speed of sound, faster than any civilian craft ever built. At this point the ship's oxidizer tank began feeding gaseous nitrous oxide to the rocket motor instead of the liquid nitrous it had delivered until now. This was as expected, and Binnie was ready for the resulting apoplectic shuddering of the craft. The ride smoothed out after about four seconds, and that was Binnie's cue to find out if all of his hard work over the last few days was going to pay off.

"More with the pitch up, Brian," Shane prompted him.

Now, while the trim stabs still had some tenuous wisps of air to dig into, Binnie thumbed the trim control way back to get the nose up to 88 degrees from the horizontal.

Shane called out the altitude the ship would coast to if Binnie shut down the rocket now. "300,000. . . . Three-two-eight." Burt Rutan, sitting beside Shane in mission control grinned. "That's three-two-eight," repeated Shane. Three hundred twenty-eight thousand feet. If the motor shut down now, the ship would coast to 100 kilometers, the X PRIZE altitude. Now all Binnie had to do to claim the prize was get back down in one piece.

"Copy that," returned Binnie.

"Three-fifty. Suggest shutdown."

Eighty-three seconds after Binnie had fired the rocket motor, he hit the switch to shut it off. He was at 213,000 feet then and still climbing, sailing upwards on momentum, absolutely straight and true, and without a hint of a roll. The new flight procedures had worked flawlessly.

The instant the howling of the engine cut off, Binnie found himself weightless, with no sense of up or down. All the tension left his body. He sailed into a black sky completely devoid of stars. He felt, he wrote later, as though he had stepped across a threshold into an entirely different plane of existence, "a realm of blessed peace and quiet."

With the boost phase of the flight behind him, Binnie could take his eyes off the display that had commanded his attention since the drop, let his hands float from the controls, and just enjoy the view. And what a view it was! Ahead was the fathomless black void of space. He almost felt as much as saw it: a "vast presence, looming and yawning through the spaceship's little windows," full of both "menace and mystery."

But when he tilted his head back, the warm reassuring earth tones of the desert and mountains below greeted him. He could see a thousand miles in every direction, taking in not just the land but a wide expanse of the deep blue Pacific Ocean as well, along with swirling brilliant white cloud cover, even entire weather patterns, and, on the horizon, a thin, gently curving band of electric blue that was the atmosphere—all that separated the Earth from black infinity. That view, Binnie said later, "is pretty special.

But the way you get there, you know, it's such a feeling. It's like somebody's taking the cymbals at the end of a symphony and going *po-WHAM!*"

As *SpaceShipOne* sailed free of Earth's bounds, Binnie hit the controls that activated what he called the angel's wings—Rutan's "feather" that would allow the ship to reenter the atmosphere safely. The ship split itself in half, with the twin tail booms hinging upward until they locked into place almost at a perpendicular angle to the rest of the ship.

Back in mission control, as the feather locked into position, Richard Branson shook Paul Allen's hand, acknowledging the handoff from Allen's spaceship development program to Branson's commercial spaceline venture. That moment marked the end of the beginning of the commercial space age. Allen in turn clapped Rutan on the back, and the two of them shook hands, too, Rutan grinning like a kid on his birthday, mission very nearly accomplished. There was just that little matter of the landing to attend to.

"You're gonna want to orient northwest for the reentry, Brian," advised Shane, unflappable as always.

"Okay, Doug, copy that, northwest."

"You sound great. You feeling good?"

"Aw, I feel great." The feeling in Binnie's voice said it all: that view and the reprieve from gravity were worth every one of the long hours, days, weeks, months, years he had worked to get here. As he sailed upward through 328,000 feet he broke out a digital camera and began snapping pictures out the windows. The ship's attitude was so stable he had to use the RCS to set the ship slowly spinning and get better views of the Earth.

"X–15," radioed Shane. This was the last milestone Rutan had hoped to reach, the 354,200-foot mark set in 1963 by the craft that was his inspiration, X–15 rocket plane. From the cockpit of the *White Knight*, Melvill radioed his congratulations.

"Outstanding," said Binnie. Then, "Boy, it's really quiet up here."

All too quickly, the spaceship reached the top of its arc: 367,500 feet, or 69.6 miles straight up. Then it surrendered to gravity and began to fall. After just three and a half minutes of weightlessness, it hit the upper fringes of the atmosphere, and the quiet, floating peace Binnie had enjoyed in space was broken. He heard the wind outside as a distant roar in the rarefied atmosphere that grew steadily louder. The g forces built up to a peak of 5 g's, then eased off. The ship bucked and shook as it slowed to subsonic speeds. Then, at 60,000 feet, Binnie brought the tail booms down, and he was back to piloting the ship, this time as an unpowered glider. He'd flown the perfect flight. But his job wasn't done yet.

Like everything else about flying the spaceship, landing it was a major test of piloting skills. Ordinary gliders have spoilers on the wings to help control the plane's rate of descent during landing. Not *SpaceShipOne*, which means, Binnie explained, "we have to fly a high, fast approach. The way that you're coming in, it looks like no way you're going to make the runway. But when you put the gear out, they're such high drag devices the thing drops like a rock, and you actually have to push over to keep the air speed up. And now you're in this 20-degree bombing run for the runway. You need to be able to flare out of that appropriately or you're going to be in trouble."

At thirteen hundred feet, Binnie pulled a lever, and the landing gear and nose skid snapped out like a trio of switchblades. Chuck Coleman, chasing *SpaceShipOne* in his Extra 300, talked Binnie down the last few feet: "One hundred, looking good. . . . That's twenty. . . . Five, start to level off. Four, keep her coming down. Looking good. Four, three, two, one. And one foot, " Coleman called out as Binnie leveled off in a picture-perfect flare. "One foot." *SpaceShipOne*'s wheels ever-so-gently kissed the runway, and "Sweetness!" Coleman exulted. "Congratulations, Brian!"

"I'm proud of you, man!" called Melvill.

"Thanks, Mike," returned Binnie.

"Outstanding job, Brian," added Shane as the ship rolled to a stop.

"Thank you, Doug. Appreciate it." Binnie's voice remained level, but he was grinning under his oxygen mask. He'd done it. He'd brought home the gold for Scaled Composites and Paul Allen.

Burt Rutan was ecstatic about his golf buddy's performance during the impromptu press conference on the tarmac after the flight. "The last thing that I said to Brian before we closed up the door around 6:00 this morning was to use a driver, keep your head down, and swing smooth." He turned to Binnie with a grin. "And I'd like to say to Brian right now, 'Nice drive.'"

When it was his turn to speak, Binnie unfurled an American flag he had taken with him to space and said, "I wake up every morning and thank God I live in a country where all of this is possible." Then he climbed on top of the ship, still holding the flag, and let himself be towed down the flight line as Melvill had done after his flights so that the crowds could see the spaceship and its pilot up close. The only way to cap a perfect day like that was to take a congratulatory call from the President of the United States. No one's grin was bigger than Binnie's in the Scaled conference room when the call came in.

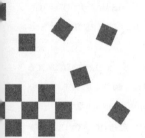

5

NASA Hitches a Ride

After the X PRIZE

The X PRIZE win had the effect Peter Diamandis intended: the world's perception of spaceflight changed forever, and an influx of cash greeted the fledgling commercial spaceflight industry—some of it from NASA itself.

Congress responded with legislation that would allow passengers to fly at their own risk until the bugs got worked out of the new suborbital tourist spaceships on the drawing boards—in effect sanctifying the Wild West aspect of commercial spaceflight. President Bush signed the Commercial Space Launch Amendments Act on December 23, 2004, giving commercial spaceship operators eight years—through 2012—to fly free of regulation from the FAA, at least as far as passengers and crew were concerned. There was just one condition: if anyone got hurt, or even had a close call on a spaceship, the government could step in and shut it down.

Lenient though it appeared to be, that was still a high standard. Mike Melvill's series of rolls as he left the atmosphere on X1 and Brian Binnie's apparent near-miss of *White Knight* on X2 might both qualify as potential show-stoppers under the new legislation. Even in the case of a perfect flight, spaceship operators in the post–X PRIZE world were strictly regulated to protect the safety of the uninvolved public from crashes or other accidents. Still, the Space Launch Amendments Act gave the upcoming commercial spacelines the room they needed to maneuver, at least for the near future.

Not everyone thought that was such a good idea, however. Representative James Oberstar, a Democrat from Minnesota, derided the act as promoting a "tombstone mentality" in which passengers would become the hapless subjects of dangerous experiments in commercial spaceflight. "Experimentation with human lives," said Oberstar; "we don't allow that in the laboratories of the Food and Drug Administration or the National Cancer Institute; why should we allow it with space travel?" Oberstar thought that commercial spaceflight should be regulated just like the airline industry.

But if they did that, argued the bill's sponsor, Representative Dana Rohrabacher, Republican from California, during debate on the bill on the floor of the House of Representatives, the new industry would be "strangled in its crib by overregulation." His feeling was echoed by Sherwood Boehlert, Republican from New York, who said on the House floor, "This industry is at the stage when it is the preserve of visionaries and daredevils and adventurers. These are people who will fly at their own risk to try out new technologies. These are people who do not expect and should not expect to be protected by the government. Such protection would only stifle innovation."

Amen to that, said the space entrepreneurs. "We are on the verge of opening up the greatest frontier that humanity has ever had," said Peter Diamandis at the International Space Development Conference (ISDC) in Washington in May 2005. "And we are sitting here saying, 'Well, someone might die.' Yes! Someone *will* die. And that's okay. It's okay for us to risk our lives for things that we believe in on the verge of the greatest expansion we've ever had."

Virgin Galactic president Will Whitehorn, also speaking at the ISDC, expressed a more measured approach to risk. "The 30,000 people who've now spoken to us about wanting to fly are choosing to do so because they believe that this is safe—and they've always wanted to go into space. Safety is really the top of people's list as to why they think they're interested in flying a suborbital spaceflight." To put it simply, Whitehorn thought no one would pay $200,000 to ride a spaceship they thought had a good chance of killing them, no matter how badly they wanted to see the black sky. Whitehorn wasn't worried about his company's ability to field a safe spaceship; with his parent company's track record in running both an airline and a railroad, he felt he had this challenge well in hand. No, what kept him awake at night in those heady post–X PRIZE days was the idea that some backyard rocket scientist could blow someone up trying to get to space before Virgin had a chance to fly. If that happened, the United States government might well rethink its hands-off approach to commercial spaceflight and ground *all* flights.

Even as the government and NewSpace, as the entrepreneurial space community began to call itself, sized each other up, one new space venture used the ISDC to bridge the gap between the two, and not just with words. Transformational Space Corporation, or t/Space, brought a full-size mock-up of its planned spaceship to the conference.

The crew transfer vehicle, or CXV, was the only spaceship at the conference, and it dominated the exhibit hall, crowding the book and video dealers and space artists exhibiting their wares to the sides of the room. The CXV took up so much space that you couldn't cross from one side of the room to the other without first swinging into orbit around its gently curving hull. Built by major Hollywood prop maker and spacesuit costumer Global Ef-

A t/Space capsule mock-up fills the exhibit hall at the 2005 International Space Development Conference. *(Courtesy of Transformational Space Corporation)*

fects, which also made prototype spacesuits for NASA, the mock-up had a certain . . . gravitas.

Like Apollo, it was a capsule and hence devoid of wings or control surfaces. But it resembled the Apollo capsule only superficially. Its nose was markedly blunter and rounder, for one thing, based on the self-righting shape of the unmanned Corona capsules that had brought back spy camera film from orbit for the United States government in the 1960s. The back end of the CXV sported a hatch and maneuvering rockets instead of an Apollo-style heat shield. Without a side-mounted hatch, as on Apollo, the ship's hull was uniformly smooth white, with faint markings to indicate the two layers of reusable heat-resistant tiles t/Space proposed in place of an Apollo-style ablative heat shield. Unlike the Apollo capsule, the CXV was designed to be flown more than once. Three to six astronauts (depending on how much gear they were taking with them) would ride to orbit on top of a booster stack consisting of two expendable stages, fueled by propane and liquid oxygen, that would be launched from a jet airplane. Coming back from space, the CXV would nose into the atmosphere, plunging down like the head of a shuttlecock. The seats inside would swivel around 180 degrees to enable the astronauts to take the deceleration from orbital velocity the same way they took the acceleration on the way up—on their backs, where it would be most comfortable. The CXV was unique not only in its design but also in the way it was funded—by NASA, which was looking for new ideas from the emerging commercial space industry.

Two events in 2004 had led NASA to seriously consider working with space entrepreneurs. One of these, of course, was Scaled's success with *SpaceShipOne*. The other was a new directive from the president of the United States. In January, even as Scaled Composites was preparing for its first spaceflight, President Bush announced to the world that he had a new mission for NASA: to send people back to the moon before 2020, and from there on to Mars.

Bush's announcement sent a shock wave through NASA, just as President Kennedy's first "we're going to the moon" announcement had back in 1961. And although Bush didn't say so, it seemed probable, given its timing, that his announcement was politically motivated, just as Kennedy's had been. Only this time it wasn't the Russians we had to beat to the moon; it was the Chinese, who had sent the first Chinese astronaut, dubbed a taikonaut, into space just the year before and who had moon-aspirations of their own.

As space analyst and former NASA engineer James Oberg put it in testimony to the Senate Committee on Commerce, Science, and Transportation, "the future of lunar exploration—and China's role in it—is likely to be extremely interesting. While the motivations that fueled the Space Race of the 1960s are largely absent ... there remain solid motives for international rivalry, for serious attempts at illicit technology transfer, and for activities that could diminish the world stature of U.S. aerospace technology."

Kennedy had couched his goal of "landing a man on the moon and returning him safely to the earth" explicitly in the context of demonstrating to the world the superiority of the United States while making oblique references to the Soviet Union, with which the United States was engaged in a cold war. Said Kennedy, "If we are to win the battle that is now going on around the world between freedom and tyranny, the dramatic achievements in space which occurred in recent weeks should have made clear to us all, as did the Sputnik in 1957, the impact of this adventure on the minds of men everywhere, who are attempting to make a determination of which road they should take."

Bush, by contrast, denied the existence of a new space race. "The vision I outline today is a journey, not a race," he insisted, "and I call on other nations to join us on this journey, in a spirit of cooperation and friendship." As for why we should go back to the moon, "We choose to explore space because doing so improves our lives, and lifts our national spirit." He also said that the new moon shot would bring more of the kind of technological breakthroughs that had come out of the first moon program—communications and scientific satellites, advances in electronics and the like—and that kids inspired by the program would study more math and science. None of which struck me as terribly convincing.

More to the point was Bush's statement "We will begin the effort quickly, using existing programs and personnel." In other words, the new moon program would keep all the same companies and constituents then working on the space shuttle and the International Space Station happily in business. As it turned out, it would use much of the same hardware, too (including the solid rocket boosters and external fuel tank that had doomed two space shuttles), thus keeping the same manufacturing centers on line. It seemed clear that at least one reason for the new moon program was to provide renewed justification for the existing infrastructure.

Regardless of Bush's true motivations for wanting to return to the moon, the real trouble with Bush's moon shot was that it was largely an unfunded mandate. It instructed NASA to get the shuttles back in orbit, keep them flying until at least 2010, complete the half-finished International Space Station, and, on top of all of that, build a new fleet of moon ships to recapture the glory days of Apollo. All without increasing NASA's annual $17 billion budget by more than a billion dollars spread out over the next five years. This paltry sum was just one fifth of the $1 billion per year *in 1970s dollars* that had been allotted for space shuttle development and that had forced NASA to make major design compromises.

Bush's mandate, the Vision for Space Exploration (or VSE, as it came to be known), seemed an impossible mission for the beleaguered space agency. Nevertheless, NASA's managers took it quite seriously. The big question was, how on Earth—or off it—were they ever going to pull it off?

The success of *SpaceShipOne* pointed to a possible solution—if NASA could capture some of that *SpaceShipOne* magic for itself. Magic, from NASA's point of view, was small entrepreneurial companies with few employees to support, low overhead, and the hunger that came with having to prove themselves—something the big boys hadn't had in decades. Fortunately for them, some of that magic dropped into their collective laps right when they needed it most—in the form of t/Space.

As they often did, NASA managers looking for ideas on how best to proceed on future projects put out a request for proposals, called in government-speak a broad agency announcement, or BAA. The winners of this particular BAA would get $3 million to produce studies, essentially technical papers, on what the space agency's new moon ships might look like. The BAA included the option to award particularly promising ideas with a second $3 million for further study. t/Space was one of the companies that responded to NASA's call in the spring of 2004.

At the time, t/Space was no more than a couple of guys with

a dream. Space visionary David Gump, a journalist by training, had cofounded LunaCorp, which in 2001 had brokered the first commercial to be shot on the International Space Station: a TV spot for RadioShack. Gump had also interested RadioShack in a venture to send the first privately built planetary rover to the moon, where it would tour the *Apollo 11* landing site and let RadioShack shoppers drive through virtual moonscapes based on real lunar imagery. Unfortunately, the cost of this mission, relying as it must on the very expensive rockets then available, proved too much for Gump and his corporate sponsors, and LunaCorp folded in 2002. Still, Gump was convinced of the viability of space travel as a commercial venture—if only a sufficiently cheap vehicle could be built.

Gump had always thought that NASA had made a mistake in not using commercial spaceships for access to space after Apollo and instead had built "a national space truck," as he called the space shuttle. Now he sensed that the time was ripe for another look at the commercialization option. Just to be sure, he ran his ideas for privatizing America's access to space past Admiral Craig Steidle, the NASA manager then in charge of realizing the VSE. He came away from his talks with Steidle and his staff even more encouraged. "They seemed very open to examining what's the best way to go back [to the moon], rather than simply assuming that things be done in a government fashion," he said later. "They certainly, all of them, had seen that big government programs were not the only way to get things done."

Looking for a partner with whom to reply to NASA's latest BAA, Gump turned to his friend and fellow space entrepreneur, Gary Hudson. Hudson was skeptical, to say the least. "You're crazy," he told Gump. "NASA will never give us any business. And even if they did, it would be agony to work with them."

Hudson had had his own ups and downs in the commercial space business, most famously with his company Rotary Rocket, which had brought Brian Binnie to Mojave, and whose prototype vehicle, the Roton, eventually found itself elevated, if not to outer space, at least to the status of monument in 2006 when it was permanently enshrined at Mojave Airport as a tribute to Mojave's aerospace pioneers.

Hudson was now at work on a venture called AirLaunch under contract with the U.S. Air Force and the Defense Advanced Research Projects Agency (DARPA) through their new FALCON program. The mission of FALCON, which stood for Force Application and Launch from Continental United States, was to build an unmanned vehicle capable of bombing targets anywhere in the world from the continental United States within a couple hours' notice. The Air Force and DARPA saw quick and affordable access to space as their ticket, so step one for FALCON was to build a demonstration system capable of placing a small satellite in orbit for $5 million or less per launch.

AirLaunch had scored half a million dollars in study money for Phase I of the FALCON program to come up with ideas for the satellite launcher and was now awaiting word on whether it would make it to Phase II—to begin building the launcher itself. Nine companies, including AirLaunch, had made the final cut, and Hudson figured he had maybe a one in three chance of landing among the winners. "Those are not great odds to bet your future on," as he told me later. While he certainly didn't relish complicating his life by tangling with NASA's notoriously difficult bureaucracy, he needed to hedge his bets on his FALCON gamble. Then, too, Hudson had known and worked with Gump for a long time. Gump had helped Hudson start AirLaunch and had been on the board of directors for Rotary Rocket. So Hudson told Gump, "What the heck; I can write a proposal."

For the first $3 million in study money, Gump and Hudson proposed to NASA that they'd produce the standard presentation slides and other informational materials that NASA's BAA requested. But if they were awarded the second $3 million, they'd do the impossible: instead of just producing more paper spaceships, they'd build actual working hardware to demonstrate their concepts. Further, if NASA liked what it saw, t/Space would build a little more in exchange for a little more funding, continuing to reach milestones on the way toward a fully functioning spaceship. If the company couldn't meet a milestone, NASA could simply walk away from the deal without losing any more money. In contrast to the usual more or less open-ended contracts awarded to the big aerospace players, t/Space would get paid only for actual working hardware. For demonstrated results, in other words. This "radical" new plan was actually how most of the business world worked; it just wasn't how NASA, formed in the heat of the Cold War when beating the Russians in space was its overriding mission, cost be damned, was set up to operate.

Hudson thought it was a great idea. But he never really expected NASA to go for it. He popped his and Gump's proposal in the mail and promptly forgot about it. In fact, he was so certain he and Gump wouldn't be taken seriously that he didn't think it worth the trouble to formalize t/Space's status as a company. That is, until he got the call from NASA one day in July 2004. "I hang up the phone," he later recalled, "and I turn to my wife, and I say, 'You're not going to believe this, but we've won!' And she turns back to me and says, 'What do you mean, *we*? We haven't even incorporated the company yet.'" Hudson and Gump raced to incorporate t/Space that afternoon; they had nowhere to deposit their first company check!

What Hudson hadn't counted on was the pressure building within NASA to change the way it did business, and fast, if it was to have any chance at all of accomplishing the mission given to it by the president. Michael Lembeck, chief of the Requirements Division at NASA's Exploration Systems Mission Directorate, the department charged with realizing the VSE, was among the NASA managers excited by the promise of lean, driven entrepreneurial companies reaching space for pocket change. To Lembeck and his colleague Garry Lyles, Exploration Systems' head engineer, t/Space looked like just what the doctor ordered. Lembeck and Lyles bumped Gump and Hudson's proposal straight to the top—to Exploration Systems chief Craig Steidle—along with a strong endorsement. "We cut t/Space loose to get out of the realm of theory and into the practical," Lembeck said later. t/Space was now riding the wave set in motion by the winning of the X PRIZE.

As *if* to underscore its mission to bring entrepreneurial thinking and flexibility to NASA, t/Space chose a former NASA astronaut as its tour guide for the CXV mock-up at the ISDC.

I'd never have guessed that the soft-spoken middle-aged man in a suit and tie showing conventioneers around the inside of the CXV was one of the world's most experienced astronauts. Jim Voss had just signed on as t/Space's engineering manager, and he'd led the design work on the only functioning hardware inside the capsule, the swiveling seats. An aerospace engineer who had flown on the space shuttle five times and had lived aboard the International Space Station for five and a half months, Voss was exceptionally well qualified to design space furniture. Beneath his quietly unassumingly demeanor, he bubbled with youthful enthusiasm for space and his work on the CXV. He was quick to give most of the credit for developing the CXV's seats to undergraduates under his direction at Auburn University, where he'd been teaching since his retirement from NASA. He had just set them on the right track with the initial inspiration, he said, and helped guide them along.

The trouble with the seats on the space shuttle, Voss had thought during his spaceflights, was that they took up so much room in an already cramped spaceship, and they were useless except for launch and reentry. What we really need, he told one of his crewmates before a launch, is something more like a hammock, made of strong enough fabric to take the g-loads of launch, that then could be wadded up and put in a locker once the ship reached orbit. Not a bad idea, his crewmate told him, but good luck getting NASA to sign off on something so simple and logical in concept. Yeah, Voss agreed, and dropped the idea.

But now t/Space had given him the chance to realize it. The seats in the CXV did look a bit like hammocks, and they swiveled for that crucial 180-degree about-face on pivots set in what was the ceiling and the floor with the CXV lying on its side in the exhibit hall. A surplus Russian pressure suit sitting in one of the seats stood for a crew member, while an empty seat in the ship invited convention goers to try them out for themselves. The CXV had another feature lacking in Apollo: a zero-g commode neatly tucked away in the nose, where the parachutes would have been in an Apollo capsule. The parachutes for the CXV would be stored in the vehicle's side and deployed at the rear.

While I was talking with Voss inside the capsule, Buzz Aldrin climbed aboard to check it out. I suddenly found myself sharing the inside of a next-generation spaceship with a member of the first crew to land on the moon and one of the longest-flying astronauts in the history of spaceflight. As the two men shook hands and greeted each other by first name, I felt that I might be witnessing a pivotal moment in space history. The moment was quieter and was getting a lot less press than the flights of *SpaceShipOne* had, but it seemed potentially no less profound. The space entrepreneurs were coming into their own, attracting the high-caliber talent they needed to succeed, and now they were starting to get the best kind of government support: money to develop their ships as they saw fit for a given mission.

"What we're proposing to NASA," Gump told me when I caught up with him at the conference, "is a type of incremental side bet." A side bet to the new, big-budget moon ship that NASA was going to contract out to major aerospace companies and that was also going to have to pull double duty as an International Space Station cargo and crew vehicle. The CXV could fill that latter role nicely, Gump said, leaving the moon ship free for its primary mission. He quickly corrected me when I referred to the CXV as a shuttle replacement. Instead, said Gump, "we are a

Soyuz replacement—a simple craft to ferry people up and back." The space shuttle was a "self-propelled space station," Gump told me. That's not at all what t/Space was about.

Dressed like Voss in a conservative well-tailored suit and tie, and even more understated in manner, Gump easily blended into the crowds of space entrepreneurs, aerospace company executives, and NASA managers and engineers at the convention. He attracted little notice—except from NASA managers, who crowded around him after he gave a presentation. He answered their questions about t/Space and its plans matter-of-factly, without a hint of the grandstanding or wild speculations that characterized the speeches of some of his fellow space entrepreneurs. Gump's manner, his presentation, his company's mission, and its goals all spoke of careful and due consideration, solid engineering principles, and a true concern for mitigating risk, both to his future space crews and to NASA's pocketbook. It was a potent brew for NASA managers anxious to get a move on with the president's mission to send people back to the moon while somehow continuing the agency's current activities without significant budget increases.

Even more potent was seeing t/Space hardware in flight.

Just days after the space conference, NASA's Lembeck rode high over the Mojave Desert, watching from a chase plane as an aircraft that looked as if it owed its inspiration more to bug life than the world of aerospace roared on jet power toward a test area with a rocket slung under its belly. It was June 2005, and a test of a major component of t/Space's proposed spaceship was underway.

The buglike ship was Burt Rutan's *Proteus*, the craft he had originally envisioned as the launcher for his own spaceship. Now it flew missions like this, drop-testing airframes for the military and other customers, like t/Space (and t/Space's customer, NASA). Chuck Coleman, the pilot who had talked Mike Melvill and Brian Binnie down to the runway after their spaceflights, piloted *Proteus*. Lembeck, riding right seat in a Grumman Tiger,

A t/Space mockup undergoes drop testing over the Mojave Desert. *(Photo: Michael Lemb)*

held his digital camera at the ready as the planes approached the test area, an especially lonely patch of desert.

The "rocket" riding under *Proteus* wasn't live, and it was only a quarter of the size of the proposed full-up spaceship. It was a dummy shape meant to test the actual ship's aerodynamic characteristics. When Coleman's copilot, Mike Alsbury, flipped the arm and release switches, it would drop like a rock, but not before testing out t/Space's new idea for air launching a rocket.

The idea was the brainstorm of Brian Binnie's old friend from their Navy days, test pilot–engineer Marti Sarigul-Klijn. The traditional method for launching a rocket from an airplane— as with *SpaceShipOne*—had the airplane flying on ahead of the rocket after the drop, at which point the rocket, still in a horizontal attitude, fired its motor and used its wings to pitch up for the run to space.

That method presented some major problems. For one thing, it meant that the rocket had to have wings with control surfaces or fins to enable it to perform the pitch-up maneuver. These structures added weight and made the ship more fragile, meaning that the rocket had to carry more fuel and had more opportunities to fly apart—"failure modes," in the parlance of engineers. The extreme pitch-up maneuver required to pull up from horizontal to vertical put additional stress on the rocket in the form of increased dynamic pressure—"Q" in engineer-speak. That meant that the ship had to be beefier, adding yet more weight, requiring more fuel, and lowering the rocket's useful payload even further.

And there was the problem that may or may not have posed a risk to *SpaceShipOne* on the X2 flight—the potential for a traditionally air-launched spaceship to hit its mother ship. Since the rocket traveled so much faster than its jet-powered carrier airplane, it would certainly catch up with it after it fired its rocket motor. And if the rocket ship happened to be pitching up at just the wrong rate, its trajectory could possibly intersect with that of the airplane. Boom.

To solve these problems, Sarigul-Klijn came up with an ingenious yet simple trapeze-and-lanyard affair that would allow the t/Space ship to perform what the company called a trapeze-lanyard air drop, or t/LAD. As *Proteus* passed over the test area on the desert floor Alsbury dropped the rocket. Sarigul-Klijn's trapeze swung down to stabilize the rocket's plunge and guide it clear of the airplane. Then, released from the trapeze, the rocket fell engine first, its nose still attached to the trapeze by the lanyard. The lanyard reeled out behind the airplane, pulling the rocket toward vertical before detaching from the airplane. Finally, a small drogue parachute attached to the rocket motor's nozzle arrested the rocket's pitch-up motion and allowed the mock-up to point straight up, nose vertical, in an ideal position from which to fire a real rocket motor.

Colman had never seen anything like it. "I was astounded

that you could get this booster to do a 90-degree turn and just kind of hang there in space," he told me afterwards. "I mean, this thing just dropped straight off. It didn't roll, yaw, anything. It was like it was elevating straight down." Straight down until it impacted the desert floor in the test area as planned, kicking up a little cloud of dust that Lembeck and the others could see from the air. This was the third and last test of the t/LAD system, and it passed with flying colors. Lembeck and his bosses back at NASA headquarters in Washington couldn't have been happier. Watching t/Space at work "was extremely rewarding," Lembeck told me. Especially since "I know how easy it is to throw out a set of viewgraph presentations and claim you're going to save the world as opposed to actually building something and making it work. To see t/Space actually do this and do it not just once but three times so they show that they didn't just get lucky, but that they actually have a system that looks like it'll work for deploying a rocket off an airplane," well, it just didn't get much better than that—not for just $6 million in study money.

Later that month in Washington, NASA's newly appointed administrator, Mike Griffin, made a startling announcement to the Space Transportation Association, a trade group made up of space launch companies. Citing the lack of competition in the aerospace industry, Griffin said that he intended to fund the development of commercial spaceships to supply the International Space Station with cargo and crew. After the ships were built, NASA would buy rides on them rather than owning them outright, leaving them free to pursue other markets as well.

Sure, the new moon ships would be capable of fulfilling those missions, said Griffin. In fact, he was going to make sure of that, since he thought NASA would be foolish to put all its bets on NewSpace. Still, he'd like to have both government ships and private ships servicing the station, just as the government now used both commercial and private airplanes for terrestrial transportation.

"Expect to see the government looking to make a deal in a

Gump (center) talks to press in front of a t/Space capsule mock-up with Peter
ndis (to his left) and New Mexico economic development secretary Rick Homans at
)5 X PRIZE Cup. *(Photo: Michael Belfiore)*

commercial sense," Griffin told the group. "Rather than issuing
a prime contract focused on process and on very detailed specifications of how to do things, look for a deal-making arrange-
ment where we tell you what it is that we want the requested
service or goods to be able to perform." Under Griffin's plan,
NASA would give some commercial spaceship provider "milestone money, progress payment money, depending on him meeting and achieving certain milestones in the development of the
bird."

In other words, Griffin was committing to Gump's incremental side bet to the big-budget moon ship for sending people and
supplies to orbit. No carte blanche contracts here. NASA would
offer these new contracts for fixed prices, and if vendors couldn't

follow through, too bad for them. "This is the way people buy things out in the world," said Griffin. "I don't go and buy a car or an airplane or pretty much anything else on the basis of 'Why don't you build me this car and tell me how much it costs when you're done?'" Such a move, if Griffin could pull it off, could radically change the way the space agency did business.

But it would have to start small. Most of NASA's budget for human spaceflight, set by Congress, was committed to the space shuttle, the International Space Station, and the new moon ships. That situation was unlikely to change any time soon. That left only a couple of hundred million dollars for Griffin to spare for his new commercial space program. "There is a line in our budget called 'ISS crew and cargo services,'" said Griffin. "It's not overly well funded right now," he admitted. "But it exists, and we plan to use that to get started on this process."

Any commercial spaceship company NASA worked with would have to have what Griffin called "skin in the game"—its own source of funds. "Despite the wishes or entreaties of those who might want me to dump $400 or $500 million on their enterprise . . . and to stand back and wait to see if the results come in, that's not going to happen," said Griffin in an oblique reference to the amount t/Space needed to build its spaceship. The government would kick in some development money and guarantee a market for a finished spaceship, but the company would have to take on much of the financial risk itself. Given NASA's limited funds for the new program, this was a practical consideration. But it was also another way for Griffin to cover his bets. Any potential space transportation provider would have to have a reasonable chance of success with or without NASA funding.

Unfortunately for Gump, "skin in the game" was just what he lacked. By August, t/Space had run through most of its NASA study money, and there was only enough left for just one more in-flight hardware test. On a clear Wednesday morning off the northern California coast, a cargo helicopter dropped a full-size CXV mock-up from ten thousand feet. The mock-up deployed three blue-and-orange parachutes and floated down toward the water while a smaller chopper circled to let a photographer snap pictures of the drop and give Gump a bird's-eye view of the test. t/Space was so low on money by then that it had to split the cost of the second helicopter with *Popular Science*, which had also sent the photographer to shoot pictures for an article I was writing.

Splash! The capsule hit the water at a gentle fifteen miles per hour, just as it would on a return from space. As the chutes settled around it a boat rumbled up and two divers jumped into the water to retrieve them for a future test. Mission accomplished. With that, NASA completed its first foray into entrepreneurial spaceflight. By any measure, the experiment was a success. The question remained for t/Space, however: would it continue to benefit from NASA's new way of doing business when Griffin opened the doors to competition?

True to Griffin's word, NASA released its formal request for proposals in January 2006. The new program, now called Commercial Orbital Transportation Services (COTS), attracted some two dozen applicants. In the meantime, Gump had managed to get some "skin" in the form of investors who would pitch in if he got another NASA contract, and t/Space emerged as one of six finalists competing for a pot now worth $500 million over the next five years. Just two entrepreneurial space companies made the final cut that summer. Unfortunately for Gump and Hudson, however, t/Space was not among them.

By then, Hudson's Phase II FALCON funding had come through to the tune of more than $29 million, and he was able to return his full attention to AirLaunch. Gump vowed to build the CXV without NASA's help. But although the boosters and launch technology for t/Space's CXV would continue development at AirLaunch, Gump had to admit that he now faced a much steeper climb to orbit.

Both of the COTS winners were already well advanced on their proposed orbital spaceships, and both had skin in the game. Internet millionaire Elon Musk powered one of these companies, called Space Exploration Technologies, or SpaceX, with the proceeds of the sale of the PayPal Internet payment service. Musk had helped to found PayPal, and he had sold it for a billion and a half dollars to online auction company eBay. Musk wanted to land people on Mars, with an ultimate goal of colonizing the red planet. Starting small, SpaceX had already built a lox-kerosene satellite launcher called the Falcon 1. Its main engine, the Merlin, would also power SpaceX's planned space capsule launcher, the Falcon 9, named for its cluster of nine Merlin engines.

Kistler Aerospace, the other company with a brand-new inaugural COTS contract, was founded by a Swiss immigrant engineer named Walter Kistler in 1993. It had gotten three quarters of the way done with building a fully reusable satellite launcher called the K–1 before going bankrupt in 2003. The company's hardware, including a clutch of Russian-made engines, was sitting in warehouses around the country. Many of its managers were retired NASA engineers and managers who were ready to go back into action when the cavalry arrived in the form of spaceship startup and former X PRIZE competitor Rocketplane. The two companies merged to form Rocketplane Kistler (RpK), just in time to apply for COTS. RpK's "skin" came from Rocketplane president and advertising entrepreneur George French, along with $18 million worth of tax credits from the state of Oklahoma, where Rocketplane was based.

NASA's endorsement of NewSpace was just one example of the surge of activity following the flights of *SpaceShipOne* and the winning of the X PRIZE. Peter Diamandis, for one, wasn't content to rest on his laurels. Now that he had helped to jump-start a new space age, he wanted to *do* something with it. Specifically, he wanted to bring it to the next level. With the X PRIZE he'd helped to establish privately funded space travel not only as a possibility but also as a potentially viable business proposition. What was next?

Diamandis decided that *next* had to be a way to take commercial space travel to the mainstream, to make it part of the fabric of life of millions of ordinary people. The day when large numbers of people would routinely rocket into space was still many years off, however, so Diamandis had to come up with a way to let people participate in the new space age before they could actually get to space themselves. They'd have to be more like spectators than participants at first.

Spectators. . . .

The light bulb went on over Diamandis' head when his friend and race car investor Granger Whitelaw invited him to attend his first Indy 500. It wasn't the race itself that excited Diamandis—he actually found cars roaring around a closed track rather boring, despite the best efforts of his friend to get him excited about it. No, it was all the economic activity Diamandis saw around him at the track that turned him on. An orgy of it, in fact. People selling T-shirts and ball caps, flags and jackets, all manner of souvenirs. The race was making money, lots of it, and not just from ticket sales. "I'm an entrepreneur," Diamandis said later, "and all around me in the hours that led up to the race all I saw was business. I saw huge business transactions. The TV cameras, the logos, and the merchandise."

Diamandis and Whitelaw stayed up late that night, hashing out ideas at an Indianapolis Steak 'n' Shake. By morning, they had the answer to Diamandis' next challenge: the Rocket Racing League. Commercial launch technology wasn't near the point yet where rockets could race each other to space. So instead of spaceships, the Rocket Racing League (RRL) would race rocketized airplanes on a three-dimensional course—up and down as well as around in circles like race cars.

Rocket racing would become the first big new spectator sport of the twenty-first century, proclaimed Diamandis and Whitelaw at a press conference in New York City in October 2005. Not only would rocketry break new ground, but so would the spectator experience, with handheld displays providing cockpit camera views of fans' favorite racers. The racers would fly through a series of virtual hoops that the pilots would see in military-style head-up displays and that fans would see superimposed on their camera views. The main attraction: airplanes based on Burt Rutan's Long-EZ design, with the rear-facing propellers replaced by eighteen-hundred-pound-thrust, liquid-fueled rocket engines blasting out bright orange ten-foot flames. The rocket planes, the X-Racers, would be purchased from a commercial airplane builder and modified according to standardized specifications by XCOR Aerospace, based in Mojave, California.

I saw the inspiration for the X-Racers in action in Las Cruces, New Mexico, a few days after the press conference. XCOR PR man Rich Pournelle slipped me through the gate at the airport and onto the tarmac, where an XCOR crew was fueling a modified Long-EZ dubbed the EZ-Rocket.

The EZ-Rocket had wowed Oshkosh Air Show spectators back in 2002, and XCOR had brought it out of retirement for a demo for the embryonic Rocket Racing League at the X PRIZE Cup, another of Diamandis' X PRIZE follow-ons. Twin lox-alcohol, four-hundred-pound-thrust rocket motors pointed out the back of the fuselage where the propeller had been. A bullet-shaped fuel tank had been added to the airplane's belly, and the airplane itself had been painted white with the checkerboard-racing-flag logo of the Rocket Racing League. "Go easy on us," XCOR engineer, office manager, and self-proclaimed company "Mom" Aleta Jackson entreated me as I joined a small group of photographers and journalists a standing few yards from the EZ-Rocket. "She's just a hangar queen." Meaning the airplane had only recently come out of storage and still had a few first-generation kinks that would be worked out in the prototype X-Racer that XCOR had been contracted to build for the RRL.

Former space shuttle commander Rick Searfoss had geared up in a blue flight suit festooned with shuttle mission and Mach–25 club patches to fly a test run before the demo two days' hence. He showed me his knee board with his flight plan drawn out on

00-pound-thrust rocket motors power XCOR's EZ-Rocket. *(Photo: Michael Belfiore)*

it. A diagram depicted the EZ-Rocket flying through a series of gentle turns over the airport, with periodic rocket boosts highlighted in bursts of yellow. Searfoss was now officially the RRL's lead pilot (other RRL pilots would include Erik Lindbergh and accomplished aerobatic pilot Sean Tucker), as well as XCOR's chief test pilot.

For Searfoss, the choice after retiring from flight status at NASA in 1998 had been clear. It was either "descend into management," as he and his fellow astronauts termed it, or leave the government space program. Searfoss, a trained Navy test pilot and former Air Force flight instructor, most certainly did not want to get stuck flying a desk for the rest of his working life. So he went into business for himself as a motivational speaker, took the occasional gig teaching students at Mojave Airport's Civilian Test Pilot School, and eventually staked his future spaceflight career on XCOR, a company he felt had the right stuff for building commercial spaceships.

XCOR's founders, refugees from Gary Hudson's Rotary Rocket, were dedicated to the proposition that space should be just another place to do business in, or to visit on vacation. In fact, both Aleta Jackson and company president Jeff Greason intended to retire there (Jackson planned to live on the moon, Greason on Mars). Searfoss, with his test pilot experience and two aeronautics degrees, was in a good position to evaluate XCOR's work with liquid-fueled rockets, and he came to the conclusion that the company was among the best in the business at building safe, reliable engines for use on manned vehicles. More than that, Searfoss saw XCOR as his ticket back to space. He intended to be in the pilot seat when XCOR's planned suborbital ship, the *Xerus*, in development since the company was founded in 1999, finally took off. With contracts from the RRL, and increasingly from NASA and the Department of Defense, XCOR had advanced from struggling startup to relative prosperity in just the year between Scaled's X PRIZE win and the launch of the Rocket Racing League. It seemed likely that Searfoss wouldn't have long to wait for his return ticket to space.

After Greason, Jackson, and XCOR chief engineer Dan De-Long had finished fueling and prepping the EZ-Rocket, Searfoss climbed into the single-seat cockpit with its bubble canopy. The airplane had only enough fuel for a couple of minutes of flight, and none to waste on taxiing, so the EZ-Rocket got towed behind a pickup truck to launch position on the runway, out of view from where I stood.

The pickup returned to the little group of spectators, and we waited along with a fire truck and firefighters standing by in case of trouble. Then, with a roar that sounded like a jet's, only without the turbofan whine, Searfoss rocketed down the runway, pushed by two blue jets of alcohol flame that were just barely visible in the New Mexico midmorning sun.

The EZ-Rocket fairly leaped off the runway, climbed rapidly, rocketed through a right turn, and then sailed away from us in silence with the engines off. Searfoss glided back toward us and turned left to glide beside the runway before restarting one of the engines for another left turn, pointing the sound at us for maximum boom as he sailed back around for another pass down the runway with the engines off. Halfway down the length of the runway, he tried for another relight, but this time the engine only popped and sputtered. No matter. He set the ship down in a dead stick landing and rolled to a stop. The pickup truck towed him back to his starting point. He was grinning on his return. It had been an almost-perfect flight, and there was still plenty of time before the demo to shake out the last of the EZ-Rocket's cobwebs.

Hangar queen or not, the EZ-Rocket was one of a kind. Mid-air restarts were unheard of for a rocket-powered aircraft, and this was one of the special features of the EZ-Rocket that made it attractive as the basis of the X-Racers. And, not incidentally, for the next generation of manned spaceships as well.

If spaceships were ever to become as common as jetliners, they would have to demonstrate airline-style safety and reliability, which might well include restartable engines. To keep to airline-style schedules, they'd also have to demonstrate rapid turnaround between flights. *SpaceShipOne*'s several-day turnaround between X PRIZE flights was a vast improvement over the space shuttles' months-long refurbishments between launches, but Diamandis wanted to reduce that time to mere *minutes* for the X-Racers and, by extension, for future spaceships.

With no more than three and a half minutes' worth of fuel on board, the X-Racer pilots would have to carefully husband their forward momentum, their "energy," during a race, applying rocket bursts for quick climbs or for carefully considered passing maneuvers. To finish an hour-and-a-half-long race, they'd have to land four or five times for race-car style pit stops. The EZ-Rocket's three-hour turnaround time, brief though it was for a rocket-powered vehicle, just wouldn't cut it for an X-Racer.

The biggest bottleneck to rapid refueling was the liquid oxygen tank. Super-cold lox hit the relatively warm tank like water on a frying pan, flashing to steam. The resulting vapor had to be vented through an escape valve until the tank cooled sufficiently, making fueling a tediously long process. Greason and his crew couldn't see around that problem when Diamandis told them he wanted the X-Racers to be refueled in under ten minutes tops, ideally in under five. "We were afraid we might have a law of physics problem," Greason recalled later. In fact, he refused to sign off on the RRL contract until he knew whether rapid lox filling was even possible. "We don't like to agree to do things we don't know can be done," he explained.

Prechilling the lox tank cut down the fill time, but not enough. So XCOR engineers had to invent a new process for getting the rest of the way there. In the end, they pulled it off, though Greason wouldn't reveal to me exactly how; that was a trade secret. Diamandis was ecstatic. "I remember getting their e-mail," he told me. "They said, 'Amazing news! We just did a rapid liquid oxygen fueling test, and it was so fast we had to do it again to prove to ourselves that we weren't dreaming.'" XCOR had just filled a tank with two hundred fifty pounds of liquid oxygen in fifty seconds—a feat that, to Greason's knowledge, no one had ever approached. The RRL was off to the races.

Ultimately, Diamandis and Whitelaw envisioned mass spectator events on the order of the Indy 500 auto races that would garner major network TV coverage and their accompanying millions of dollars in corporate sponsorships. And just as auto racing had fostered advances in automobile technology like radial tires, disk brakes, seat belts, and even rear-view mirrors, so too, Diamandis hoped, would the Rocket Racing League spur improvements in rocket design that could carry over to spaceships. Rapid turnaround between flights was only the beginning. Perhaps the biggest benefit to commercial spaceflight would come from the introduction of standardized parts to the industry.

The reason building rockets was so prohibitively expensive, Greason explained to me, was not because of any technical challenges, which he felt were well in hand, but because rocket builders couldn't just buy parts at their local rocket supply outlet. Those outlets and the parts to fill them with simply didn't exist. "If you sat down to build a car," said Greason, "and nobody was in the business of selling engines, and nobody sold tires, and nobody sold shocks, and nobody sold drive trains, and every single one of these pieces had to be designed and built and tested and debugged and fabricated by you for your unique requirement here to have a car, I'm not sure what that would cost, but it would be in the many, many, many million-dollar range." That was the situation faced by rocketeers at the beginning of the twenty-first century. But with as many as ten X-Racers entering races run by a successful Rocket Racing League in the next decade, that situation could change as entrepreneurs manufactured and sold standardized rocket parts that would be available to anyone who wanted to build a rocket, not just RRL teams.

Of course, all this depended on that most fickle of variables: the taste of the general public for a new sporting event that would have to compete with the likes of NASCAR and Indy-Car. Diamandis and Whitelaw were counting on the spectacle rocket planes thundering overhead, swooping down for pit stops, blasting back up again, passing each other, and blowing past the stands to draw viewers as well as video gamers who would fly virtual X-Racers against the real ones. They figured that rocket racing, with its roar of rocket engines and long flaming exhaust trails, would generate even more excitement than auto racing.

But would it? Racing fans love auto racing for the risks that race car drivers take, IndyCar executive vice president Fred Nation told me when I asked him about auto racing's appeal. The drivers are close enough to swap paint, creating a palpable sense of danger. "You can clearly identify with the risks that auto racing drivers take because you can see that it's close," Nation explained. Not to mention the fiery crashes that result when drivers miscalculate or their cars suffer a mechanical failure, which adds that much more spice to the affair.

The X-Racers, by contrast, would launch for each race in a

staggered configuration, timed to reduce the risk of midair collisions and to keep at least some of those pit-stopping racers in the air at all times, and the race would be run with wide virtual lanes between the airplanes. There would be no paint swapped here. And at maximum speeds of around 230 miles per hour, these planes would be slow by rocket standards. Slower even than the old-fashioned piston-powered race airplanes flown at the annual Reno Air Races, which have yet to match the Indy 500's popularity.

Nevertheless, the crowd lining the fences along the runway at the first annual X PRIZE Cup expo certainly loved the EZ-Rocket, especially since it was the only manned rocket vehicle to make an appearance. "Let me tell you," Searfoss told the rocket fans afterwards, "it's a kick in the pants." Akin, he told me separately, to flying a fighter jet. The X-Racers' high thrust-to-weight ratio would give their pilots quite a jolt of acceleration during their five- to thirty-second rocket boosts.

Rocket racing as a spectator sport. Suborbital tourist flights. NASA funding for space startups. These were just the beginning of the post–X PRIZE boom. "I see innovation cycles coming on top of each other," Rutan said shortly after winning the X PRIZE. "One for suborbital manned spaceflight, one for orbital, and one for going to the moon and the planets." His own ultimate goal: "affordable travel to the moon in my lifetime." In other words, the best was yet to come.

6

THE 200-G ROLLER COASTER

Rocketplane, Virgin Galactic, and the First Commercial Spacelines

Chuck Lauer was finally getting his day. He'd always thought that tourist flights to space made good business sense, even when it was unfashionable to do so. Back in mid–1995, when he'd cofounded Rocketplane, the prevailing wisdom among rocketeers was that the real money to be made was in satellite launchers. "I used to get beat up, down, and sideways by my board of directors for even talking about space tourism," he said later. No one would take the company seriously, his board told him, if one of its principals went around talking about sending people into space for fun. Then, in the late 1990s, the bottom dropped out of the satellite launch market. In 2001, the Russian Space Agency sent the first paying passenger into space aboard a Soyuz space capsule, proving there was a market for space tourism, and then *SpaceShipOne* came along and showed that a private company could make it work. Suddenly all anyone was talking about was space tourism.

Lauer could be forgiven for radiating a certain air of smugness at space conferences following the X PRIZE win. Tall, with a regal bearing, a mane of wavy white hair, and a neatly trimmed beard, he was hard to miss. His manner was as bold as his appearance: friendly, yet forceful. By then Rocketplane's business development manager, he lived to counter objections, just as he had during the past twenty-five years as a Michigan real estate developer.

He'd been infected with space fever in the late 1970s as a graduate student in architecture and urban planning at the University of Michigan. Searching for a topic for his master's thesis, he'd stumbled on the work of Gerard O'Neill and was instantly seized with the possibilities of developing outer space. As he recalled, "the high frontier vision of extraterrestrial resource development, large spin-gravity, city-sized space habitats, and gigawatt-scale space solar power made from lunar and asteroidal resources were all things that had a big resonance for me." He wrote a thesis called "The Humanization of Space" and decided to make space development his life's work. But, "I figured that I needed first of all to know how to make a deal and secondly make some money." Ten years later he helped found Rocketplane. Ten years after that, in April 2005, I found myself having dinner with him at an Oklahoma City sushi bar, along with Rocketplane's Japanese business representative and Rocketplane's first customer, sixty-six-year-old Reda Anderson.

Anderson stabbed a finger at Lauer and said, "You have one year to find me a man."

"Me?" laughed Lauer. "You have to do that."

"I can't do that," said Anderson. "I've tried."

Lauer had just finished describing the marketing scheme he and our other tablemate, Ms. Misuzu Onuki, had hatched together: Rocketplane Kistler would host the first outer space marriages.

Rocketplane Kistler's *Rocketplane XP* would be a suborbital vehicle, like *SpaceShipOne*, imparting only four minutes of weightlessness after its rocket engine cut off and it coasted out of the atmosphere at supersonic speed. The bride and groom would have to work fast, and in cramped quarters. Nevertheless, Misuzu had already been collaborating with a fashion designer back home in Japan on a 0-g wedding dress whose white trusses would rise in graceful undulating ripples below the bride's seatbelts when weightless, like the folds of a sea anemone. The ship would have four seats. The pilot would have his hands full flying it. The bride and groom would ride in the back. That left the right front seat free for a priest, rabbi, or some adventurous justice of the peace: "Do you take this man to be your outer-space-wedded husband?"

"I do," says the bride, and kisses the groom a hundred klicks over the Oklahoma plains just as the terminator between night and day touches the Rocky Mountains and sets them alight in glowing oranges and yellows like molten metal, and eternity stares down on them from the deepest, darkest sky imaginable.

Much as Anderson liked the idea, she lacked a crucial ingredient: a groom. Since she was slated to become the first paying passenger aboard the *Rocketplane XP,* she needed to find one fast. She was up for the challenge; not only would getting married in space make the perfect cap to a once-in-a-lifetime experience, but it would serve as the ideal test of her intended mate. He'd have to be someone rich enough to afford the trip, and he'd have to have an adventurous soul.

Anderson doesn't look anywhere near her age. Slim and athletic, her straw-brown hair cut boyishly close, she walks with a firm, determined stride. She smiles easily and laughs often, but a steely glint in her eye hints at the no-holds-barred deal making that earned her a minor fortune in California real estate. She

A Rocketplane mock-up emerges from the fog at the 2005 X PRIZE Cup. (*Photo: Michael Belfiore*)

worked hard for her money, and now she plays just as hard. No serene ocean cruising for her; she's as likely to ride under the waves as above them, the way she did when she dove to the bottom of the Atlantic in a Russian submersible to see the wreck of the *Titanic* firsthand.

After watching *SpaceShipOne* rocket into space, she just couldn't resist the idea of going herself. She met Lauer at a space conference only days after the X PRIZE win, and after Lauer gave her the Rocketplane pitch, she asked for one of his business cards. She wrote "Number one signed customer" on the back of it, signed it, wrapped a dollar bill around it, and handed it back to Lauer, who was stunned speechless. He realized Anderson had just handed him a bona fide contract. "There are four elements of a contract," Lauer said later. "You have a meeting of the minds of the parties, you have a known product, you have a deliverable, and you have what's called consideration, which is what the dollar bill was for." The two real estate wheeler-dealers were an ideal business match; they spoke the same language.

Now that SpaceShipOne had won the X PRIZE, a new space race was on. The winner would claim the prestige and market share that came with rocketing the first paying passengers into suborbital space. Although Diamandis and Maryniak had designed the X PRIZE with the idea that the winner would have the first market-ready spaceliner, X2 was SpaceShipOne's last spaceflight. Burt Rutan wanted his not-yet-built SpaceShipTwo to carry the first passengers. And that gave other rocketeers a chance to catch up to his lead.

On September 27, 2004, a typically cool and gray fall day in London, Sir Richard Branson held a press conference. Although deep in preparations for the upcoming X PRIZE flights, Rutan flew out to meet him at the Royal Aeronautical Society in the West End. Together they announced what they planned as the world's first spaceline. Branson already owned a successful airline, Virgin Atlantic Airways, and many other companies sporting the Virgin logo. But his Virgin Galactic would be the first to send the Virgin brand into the final frontier. Scaled Composites would build a fleet of five suborbital tourist ships, which could begin service as early as 2007. Price per ticket: £115,000, or $190,000.

Back in Mojave for the X PRIZE flights, Branson joined Rutan and Allen in mission control as Melvill and Binnie cleared the atmosphere. SpaceShipOne now carried the bright red Virgin logo on its tail booms and rocket fairing for all the world's news media to see. Rutan's feat of engineering was no longer just a technological stunt; it was the beginning of a new industry.

None of which had been on Branson's mind when he founded his first business, a music magazine, while still in boarding school in England, although he *had* been inspired by space as a teenager watching the science fiction movie *Barbaralla*. Branson's Virgin Records, which he launched in the early 1970s when he was twenty-one years old, and so called because he was a self-proclaimed novice at business, made him a killing and allowed him to expand into other businesses. That opened the way for his entry into the airline industry in the 1980s After that, getting into the space travel business didn't seem so far fetched after all.

From the beginning, Branson played by his own rules. In 1984 he horrified his managers, who thought of him as mainly a music executive, when he bought British Atlantic Airways and renamed it Virgin Atlantic Airways. By the 2000s, Virgin planes sporting painted figureheads of seminude women inspired by those on sailing ships of old, flew transoceanic routes to the Americas, Asia, and all points between. Branson's hundreds of ventures have included retail stores selling books and music, a publishing company, a cell phone network, and even his own reality TV show, *Rebel Billionaire*. Good looks, an infectious grin, and an unflagging spirit have helped make him one of Britain's best-loved celebrities, and he relishes the attention.

But one activity had always sparked Branson's imagination like no other—extreme travel. In 1986 he broke the world record for the fastest boat crossing of the Atlantic. Not content with that, he crossed the Atlantic again the following year, this time in a hot-air balloon—the first time anyone had ever done so. In 1991 he followed that up with the first balloon crossing of the Pacific.

As early as his entry into the airline business, Branson had thought about taking Virgin even higher. When he registered Virgin as a trademark for his airline, he registered it for space travel as well as for air travel. He trademarked the name Virgin Galactic in 1996 and then waited for a ship to emerge that was worthy of the name. That day came in 2003 when Virgin executive Will Whitehorn visited Scaled Composites to check on the progress of *Virgin Atlantic Global Flyer*. Rutan was building *Global Flyer* for a Branson-funded solo flight around the world by aviator Steve Fossett, and he couldn't resist showing off *SpaceShipOne* while he was at it. Whitehorn recognized it immediately as the opportunity Branson had been waiting for. "Fuck the *Global Flyer*," he exclaimed to Branson over the phone. "He's building a spaceship!"

At first, Rutan played hard to get. "*SpaceShipOne* was never intended to be anything but a research program," he said when Branson asked about flying passengers to space. Well, then, Branson wanted to know, could Rutan build him a commercial version of the spaceship? "We can't," Rutan said flatly. "We're too busy, we don't have the people, and in general we're not production people anyway; we're just prototypers." Still, Branson persisted, and finally Rutan sent him to Paul Allen, who then cut the deal that made Virgin Galactic a reality.

Branson's fleet of five spaceships, each carrying six paying passengers in addition to two pilots, would boom out of the atmosphere above Mojave several times a week within the next few years and then shift their home base to a brand-new commercial spaceport to be built near Truth or Consequences, New Mexico. The passengers would enjoy three days of preflight training to acclimate them to the g-loads and weightlessness of spaceflight. They'd also get rides on *SpaceShipTwo*'s mother ship, a Boeing 737–sized *White Knight* called *Eve* (after Branson's mother) as passengers in line for spaceflights ahead of them dropped off and blasted heavenward. As with the *White Knight* and *SpaceShipOne*, *Eve* and *SpaceShipTwo* would share identical cabin layouts, so that riding in the mother ship would also help passengers train for the ride to space.

Besides allowing Virgin Galactic to accommodate more passengers per flight, the new spaceship's roomier cabin would offer another perk not available aboard *SpaceShipOne*: at the top of the ship's arc, as it sailed out of the atmosphere, passengers would be able to release their restraints to float free of their seats. Prior training aboard *Eve* would give them the skills they needed to make it back to their seats in time for the descent back into the atmosphere and the reentry with its accompanying 4 g's of deceleration.

Eve and *SpaceShipTwo* would fly a different flight profile,

Virgin chief Richard Branson (left) and Alan Watts, the first frequent flyer to earn space travel, try out the seating in a *SpaceShipTwo* interior mock-up unveiled in New York in September 2006. *(Photo: Michael Belfiore)*

too. Instead of spiraling straight up over Mojave in the climb to launch altitude, as the *White Knight* and *SpaceShipOne* had done, the new ships would take off over land to a launch point over the Pacific ocean. After turning around to face back toward land, the mother ship would release *SpaceShipTwo* to fire its rocket engine, so that by the time it left the atmosphere it would be above the San Joaquin Valley just to the north of Bakersfield, California. After sailing through the top of its arc, the spaceship would re-enter north of the town of Tehachapi, some twenty miles to the northwest of Mojave, and glide back to a landing where it and the mother ship had taken off at Mojave Spaceport. Altogether, the spaceship would travel two hundred miles downrange, from ignition to touchdown, giving passengers stellar views of the Pacific Ocean, the hills and valleys of the coast, the Tehachapi Mountains, and the Mojave Desert on the way down.

A set of commercial astronaut wings would be pinned to the passengers' flight suits on their return, and the parties afterwards would be out of this world. On the first flight would be Branson himself, Burt Rutan, Branson's father (who by then would be over ninety), and assorted other close friends and family members. By early 2006, even before the first *SpaceShipTwo*, called *VSS Enterprise*, was completed, Virgin Galactic had collected $13 million in deposits for tickets now costing $200,000 from people like actress Victoria Principal and film director Bryan Singer (who featured Branson in a cameo role as a spaceship pilot in *Superman Returns*).

Rutan didn't think much of Branson's deal with the state of New Mexico to build Virgin Galactic's home port there. Called Spaceport America, the new facility would be constructed to Virgin's specifications, with a dedicated passenger terminal, training facilities, and resort-style lodging. The governor of New Mexico, Bill Richardson, had approved a deal for $225 million in state, local, and federal money to build the spaceport. Branson readily agreed to locate there and to begin leasing the facility for at least the next twenty years as soon as it was completed. In stark contrast, it was all Mojave Airport general manager Stu Witt could do to convince the California state legislature to approve an $11 million *loan* to build a new spaceship hangar and a passenger training center for his facility. Still, Rutan told a group of Mojave-area high school students in April 2006, "Some people have read the papers and think we're all moving from Mojave to New Mexico. That's not true at all. I have no intention of going to New Mexico; I don't think it'd be a very good place to do a spaceflight. I believe when I get out of the atmosphere I want to see the oceans and the mountains, not just the kind of crap you can see from New Mexico."

If that sounded less than charitable to the customer that had made Rutan's current work possible, it may be because some tension had developed between Rutan and Virgin Galactic. In his talk in Mojave, Rutan declined to even mention Virgin Galactic by name, referring to the company merely as "one of the

Tim Pickens blasts off on his rocket powered bicycle. *Photo: Jan Pickens Renegar.*

Brian Binnie "rides the bull" after his X PRIZE-winning spaceflight.
Photo: Michael Belfiore.

TOP LEFT: *SpaceShipOne* and *White Knight* taxi for takeoff on the first X PRIZE flight. *Photo: Michael Belfiore.*

BOTTOM LEFT: The view from *SpaceShipOne*. *Photo: Brian Binnie.*

A t/Space capsule mock-up splashes d[own] during a parachute deployment test in August 2005. *Photo: David Gump.*

R's EZ-Rocket taxis for a test flight before the 2005 X PRIZE Cup. Photo: *Michael Belfiore.*

The EZ-Rocket powers through a turn during a demonstration flight at 2005 X PRIZE Cup. *Photo: Michael Belfiore.*

The Armadillo Aerospace crew prepares for a demo flight at the 2005 X PRIZE Cup. *Photo: Michael Belfiore.*

Brian Feeney and his *Wildfire* mockup in October 2003. *Photo: Michael Belfiore.*

Blue Origin's *Goddard* test vehicle prepares for launch in November 2006. *Photo: Courtesy of Blue Origin.*

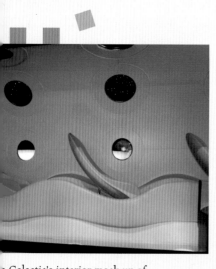

n Galactic's interior mock-up of
ShipTwo as it appeared in September 2006.
Michael Belfiore.

SpaceX's Falcon 1 rocket lifts off from the Kwajalein Atoll on its first flight in March 2006. *Photo: Thom Rogers/Space Exploration Technologies.*

A Merlin rocket engine fi[...]
at the SpaceX test facility [...]
Texas. *Photo: Thom Rogers/*
Space Exploration Technolog[...]

Bigelow Aerospace's *Genes[...]*
I test module looks down
from orbit. *Photo: Courtesy*
Bigelow Aerospace.

spacelines." One of many such outfits, in other words, that Rutan planned to supply with spaceships, and perhaps not due special consideration. One source of conflict was Virgin's need for publicity with which to sell space rides. Rutan preferred to keep quiet about work in progress. "We are back in hiding," Rutan told the students, "like we normally are. Occasionally you'll see some promotional stuff coming out of one of the spacelines, but we in general don't feel that's the right thing to do. So don't expect us to be doing any announcements or promotions or inviting the press in to look at our progress and so on. We feel it's best to let our competition think that we've quit. You just get a lot more fun showing somebody stuff that they don't expect. I will not talk about the schedule of our program, because if I get late I have to hunt up all those people and tell them why I'm late."

Rutan's closed-shop policy and, by contrast, Rocketplane's open doors were part of the reason Reda Anderson decided to put her money on Rocketplane rather than on the more heavily publicized Virgin Galactic, even though both would charge the same for comparable experiences. At Rocketplane, Anderson could quiz engineers at length about the choices they made during the design process. In fact, as Passenger Number One, she took an active role in helping make some of those choices. The engineers began to ask *her* questions about just exactly what kind of experience she wanted.

Anderson made her money in real estate rather than some intangible like stocks or gold futures for similar reasons. "Every single thing I tried to put my money in to diversify, I lost money. I said, 'You know what? Obviously if I can't see it, if I can't feel it, if I can't touch it, if I can't *smell* it, it doesn't exist for me. I'm going back to real estate.'" At Rocketplane Kistler, Anderson could thump on test hardware and demand answers about the risks involved. "If you think I don't know that I could absolutely die in this," she said, "you've got to be an idiot. It's not serious injury here; you're talking dead."

So one day in April 2006 found Anderson peering down the

throat of a sawed-off Learjet 25 fuselage at Rocketplane Kistler's workshop at the airport in Guthrie, Oklahoma. Structural engineer Derrick Seys pointed to guidelines marked on the white hull like those drawn on a patient's skin before surgery. He explained how his team would splice in part of another salvaged fuselage to lengthen the original by a good twenty inches—space needed for kerosene and liquid oxygen tanks that would power a 36,000-pound-thrust rocket engine in the plane's tail. The tail itself would be modified to replace the Lear's standard horizontal stabilizers with a V-tail that would better enable the plane's nose to pitch up coming off the runway with the heavy load of fuel. A new delta-shaped wing assembly under construction next to the fuselage would replace the Learjet's wings with a configuration optimized for supersonic flight, and, like the original wings, it would hold jet fuel.

There wouldn't be much left of the Learjet by the time workers finished transforming this gutted shell into the *Rocketplane XP:* just that fuselage, or rather two fuselages, and the Learjet 25's standard pair of General Electric CJ610 jet engines. The engines would power this hybrid spaceship to a launch altitude of twenty-five thousand feet. There, the pilot would shut down the jets and light the rocket engine for a seventy-second, 4-g boost to space and a maximum speed of three and a half times the speed of sound. After the rocket engine shut down, Anderson would get her four minutes of weightlessness, a view extending as far south as the Gulf of Mexico and west to the Rocky Mountains to take in that otherworldly sunrise, and, with any luck, her wedding-day kiss. A personal display for each passenger would let everyone toggle through views piped in from seven camera positions onboard the ship. For good measure, the pilot would use the ship's RCS to roll the planet Earth past all the windows.

Lacking *SpaceShipOne*'s carefree reentry system, the *Rocketplane XP* would have to use computerized flight controls like those on the X–15 to enable the pilot to safely navigate the dynamic pressure and supersonic speeds of reentry. Twenty-first-

A Learjet fuselage gets an extreme makeover at the Rocketplane shop in April 2006. (P
Michael Belfiore)

century computers would presumably do a better job of keeping
this ship on course than the X–15's 1960s-era electronics. And
they would give the *Rocketplane XP* at least one advantage over
SpaceShipOne. The RCS would interact smoothly with the airplane's aerodynamic control surfaces all the way up to space,
through its flight path outside the atmosphere, and on the way
back down to give the *Rocketplane XP*'s pilot complete seamless
control of the ship no matter how much or how little air it happened to be surrounded by at any given moment. That is, the
pilot would experience seamless control when he *had* control;
as with the space shuttle, the *XP*'s computers would fly the ship
from boost to reentry, with the pilot taking over only in an emergency and for landings. The pilot would restart the jet engines
at twenty thousand feet to make a powered landing. Total flight
time, from wheels up to wheels stop: one hour.

Two used business jet fuselages bolted together. A new unproven liquid-fuel rocket engine. Fly-by-wire flight controls that had to work perfectly if the craft was to survive reentry—all put together on the budget of a space startup. At first glance, it struck me as a mad scheme, and unreasonably dangerous. And yet, there was Reda Anderson, entrusting her life to it. She knew that riding the thing would be a risk, sure, but she wasn't suicidal. Much of her confidence came from the caliber of the people bold (or crazy) enough to attempt to rocketize a Learjet and blast it into space with high net-worth people on board. They were engineers recruited from NASA and from such staid and well-regarded aerospace companies as Lockheed Martin, Cessna, and the maker of those Learjet fuselages, Bombardier Aerospace. Maybe, I thought, after meeting Anderson and the Rocketplane Kistler team, the idea wasn't so crazy after all.

The idea was hatched by Mitchell Burnside Clapp, an aerospace engineer and former test pilot instructor for the Air Force, as a way to get himself to space. "Basically we're all whores for the ride," Clapp explained.

While still in the Air Force, Burnside Clapp learned that his personality was "profoundly incompatible" with life as a NASA astronaut. Mavericks and smart-asses need not apply, as he discovered early in the interview process. The clincher came when a NASA interviewer asked him: "What is your greatest weakness?" To which Burnside Clapp replied, in an attempt to add some levity to the proceedings, "Kryptonite."

"It was as if I had said a word in a foreign language," Burnside Clapp recalled. The interviewer didn't crack a smile, and Burnside Clapp knew then and there he was sunk. Screw it, he thought. If those folks couldn't take a joke, he didn't want any part of them either. So he decided to try to resurrect a perennial Air Force dream: building a manned spaceship the armed forces could call their own. His initial idea was for a single-seat rocketplane, more of a fighter plane than a passenger ship, one that could rocket into orbit to launch small satellites. Burnside Clapp called it the Black Horse in acknowledgement of its maverick nature and his hope that it would win funding against the odds.

Before he conceived of Black Horse, Burnside Clapp worked on the short-lived Delta Clipper Experimental (DC-X) single-stage-to-orbit demonstrator project. For about $60 million from the Department of Defense, a small team of engineers built an unmanned 1/3-scale model of a spaceship concept and flew it by remote control at the White Sands Missile Range in New Mexico. It looked like a science fiction rocket of old: an aerodynamic cone about four stories tall, without the multiple stages that more typical rocket ships jettisoned on the way to space. In test flights in 1993, 1994, and 1995, the DC-X blasted off; hovered several thousand feet in the air; flew sideways, still oriented vertically; and then descended to a gentle landing a quarter of a mile away. It successfully demonstrated many of the technologies needed for a full-up single-stage-to-orbit spaceship its designers hoped would soon replace the space shuttle. Scaled Composites built the DC-X's airframe. Pete Conrad, the third man to walk on the moon, remote piloted it, with Burnside Clapp serving as backup pilot.

NASA acquired the DC-X program soon after it began flying, with the intention of taking the technologies it demonstrated all the way to orbit. The result was the NASA-funded and Lockheed Martin–designed X–33 suborbital spaceship, itself a subscale version of a planned full-sized orbital ship called *VentureStar*. Since the *VentureStar* would have been owned by Lockheed Martin and leased by NASA, it was to have been the first commercial spaceship. But the technical challenges proved insurmountable, and NASA cancelled the program in 2001 after spending a billion dollars on it.

Burnside Clapp, meanwhile, had scored $100,000 from the Air Force to complete a design study of the Black Horse. When no further money was forthcoming, he left the Air Force to pursue a commercial version of the ship, reconceived by fellow aerospace engineer Robert Zubrin as a suborbital launcher of satellites.

When the single pilot reached space, he would release the satellite and an attached booster from a small cargo bay behind the cockpit. As the spaceplane sailed through the top of its arc and headed back down for reentry, the satellite would fire its booster and rocket the rest of the way to orbit. The spaceplane would, in effect, be the first stage of a satellite launching system.

Burnside Clapp got together with Zubrin, best known for selling NASA on the concept of having manned expeditions to Mars manufacture their own return fuel from elemental components of the Martian atmosphere, and Chuck Lauer to form Pioneer Rocketplane in 1996. Pioneer Rocketplane set its sights on the X PRIZE, but it struggled through the design of a larger passenger-carrying version of its spaceship chronically short of funds. Zubrin left the company in 1998, and Burnside Clapp and Lauer soldiered on alone.

Pioneer Rocketplane still hadn't left the ground by the time Scaled Composites won the X PRIZE. But it had gotten a fresh cash infusion from a new president, Wisconsin outdoor advertising businessman and space enthusiast George French, who had been an early investor in the company. As 2003 drew to a close, French, Lauer, and Burnside Clapp saw within their reach a cash prize worth even more than the X PRIZE.

They called it the O Prize. Far less publicized than the X PRIZE, the O Prize was nevertheless to prove as important to Rocketplane as the X PRIZE was to Scaled and Virgin.

The state of Oklahoma needed good jobs; young graduates who couldn't find work in the state had a way of leaving to seek work elsewhere. What Oklahoma needed was something to keep those kids close to home. The solution: tax credits to provide incentives for high-tech companies to locate their businesses in the state—businesses like Lockheed Martin's *VentureStar* project. Lockheed needed a home base for the proposed new spaceship, and in the late 1990s, Oklahoma appeared to have a good shot at attracting it, along with all the millions of dollars of investment and hundreds of new jobs it might bring. Oklahoma state senator Gilmer Capps said later, "We thought if we could get one company to come to Oklahoma, then others would follow." The Oklahoma state legislature passed Capps's tax incentive bill aimed at Lockheed in late 1999, along with legislation creating the Oklahoma Space Industry Development Authority (OSIDA). But when the X–33 got axed, it left the state of Oklahoma "all dressed up and nobody to go to the party with," as Lauer put it later. So state officials went looking for another date to the dance. Rocketplane was among those they courted.

The Oklahoma Space Industry Tax Incentive was worth $18 million in tax credits. The beauty of the credits to a startup like Rocketplane was that they were transferable, meaning that they could be sold for free and clear operating cash. To win the O Prize, a space launch company would have to be headquartered in Oklahoma, have at least $10 million already invested in it, and qualify for Oklahoma's Quality Jobs Act to ensure that the deal really would result in new jobs for the state. Like the X PRIZE, the O Prize had a deadline: it had to be won before 2004. Rocketplane beat out its competitors to win the prize in the final hour at 4:42 p.m. on December 31, 2003. French then sold the credits for $13 million, and at last the *Rocketplane XP* had wings.

Early in 2004, French brought in aerospace engineer David Urie to lead the Rocketplane design team. Burnside Clapp quit the company he had cofounded not long afterwards. "Citing creative differences is the standard Hollywood way to say that, right?" Burnside Clapp later told me. "But more specifically, they put a process together for developing the next vehicle, the next aircraft that I didn't think was going to be effective because it had way too much in common with the way they were developing the first one, and I wasn't happy with that one either." He declined to elaborate further except to say that he was skeptical of the idea that flying tourists in space was a viable business operation for something as cash intensive as building and flying a spaceship.

French would have been hard pressed to come up with a better choice as engineering chief than David Urie. Urie came to Rocketplane after thirty years' experience as an engineer and manager at Lockheed Martin's famed Skunk Works, where, not incidentally, he headed the X–33 project. During his fifty years in the aerospace industry he worked on thirty-five different flying machines, from long-haul bombers to missiles, at companies like Boeing and Douglas Missiles (before it became part of aerospace giant McDonnell Douglas). According to him, seventeen of the designs he worked on actually flew, and nine of them were still in service in the mid–2000s.

Urie couldn't resist coming out of retirement to work on one last bold aerospace engineering project. It was the business case that hooked him. "It's a way to get a small company started in a very sophisticated, high-technology area based on a market that's too small to interest the big guys," Urie told me. The chance to fly into market under the radars of monolithic aerospace companies like Urie's previous employer was just too good to pass up.

He was modest about his role at Rocketplane Kistler when we spoke at his office in the spring of 2006. "I'm just sort of a chief coordinator. My job is to develop engineers and managers and coach them and to create an organization, a company, that's capable of doing this immediate project." A jocular sign on his office door hinted at a deeper truth. "No one gets in to see the wizard," it said. "Not no one, not no how." "Wizard" certainly seemed closer to Urie's actual function than "coordinator," at least as far as some of his engineers at Rocketplane were concerned.

Once settled at Rocketplane's new headquarters in a single-story building at Oklahoma City's Will Rogers Airport, Urie set about hiring a team of engineers composed of equal parts seasoned veterans and young engineers right out of school. The latter group, in particular, looked up to Urie as a mentor as well as manager. They saw Rocketplane as a rare opportunity not only to work on an exciting project in which they would have real input on major design choices, but also to learn from one of the grand old men of aerospace engineering. True to Senator Capps's intention, many of these young engineers were Oklahoma natives who were also delighted to be able to find good jobs in their home state.

One of Urie's prime hires was engineer Bob Seto, who headed the Learjet 40 development program among other projects at Bombardier. He took over the immediate, day-to-day operation of building the *Rocketplane XP*. Talking to him made the whole enterprise actually sound almost reasonable to me. It made perfect sense to start with an existing fuselage for the spaceplane rather than trying to design one from scratch, Seto said. "There's a big cost to designing the details of a fuselage. Purchasing the fuselage reduced a lot of that risk and development effort. We don't have to spend a large amount of time starting from a blank sheet of paper."

According to Learjet's manager of technical engineering, Mike Klemanovic, that particular fuselage (Burnside Clapp's choice from the start) made a good basis for a spaceship. "The original Learjets—20 series, 23, and then evolving to 24 and 25—were based on a Swiss fighter-bomber. It's a stout, robust airplane." Bill Lear's business jet kept many of the characteristics of its fighter plane heritage, including the ability to take 3-plus g's without breaking up, and an operational ceiling around fifty thousand feet—above 90 percent of the atmosphere, with only a pound and a half of pressure per square inch separating it from vacuum. In fact, said Klemanovic, "I see that structure as not a whole lot different than a lot of previous space craft; it just didn't start out life that way, is all." With the Learjet 25's plus–3 g's tolerance rating and pressurized fuselage, the *Rocketplane XP* was already most of the way to space. Once the stock wings and tail came off, the new delta-wing assembly and V-tail would give the ship the extra structural hardiness it needed for the 4-g spaceflight.

At a maximum velocity of three to four times the speed of sound, the *Rocketplane XP* wouldn't experience anywhere near the heating from atmospheric friction that the space shuttle, traveling twenty-five times the speed of sound, encounters returning from orbit. Nevertheless, reentry heat *would* pose a problem for an ordinary Learjet's aluminum structure. Aluminum, used for airframes because of its light weight, melts at a lower temperature than a heavier metal like steel. So Rocketplane Kistler's engineers and machinists would replace the areas subjected to the greatest heat—the engine inlets, the nose, and the leading edges of the delta wing—with steel or titanium. The rest of the ship would get a coating of a special heat-dissipating paint that had been developed at NASA's Ames Research Center for next-generation spaceships and that had then been released for commercial use. Even though the bulk of the ship would still be of lightweight aluminum, all those modifications, not to mention the addition of a completely new propulsion system (the rocket engine and its fuel), added up to a much heftier craft than the original Learjet 25. The *Rocketplane XP* would top out at 19,500 pounds at takeoff, compared with the Learjet 25's fifteen-thousand-pound weight. The spaceship would need a hell of a long runway to get airborne. Fortunately, the state of Oklahoma had one.

Burns Flat, Oklahoma, is eighty miles, or a forty—minute small airplane ride, from the Will Rogers Airport. Bob Seto flew me there, along with Reda Anderson and Misuzu Onuki, in his Cessna 182. He banked on approach so that we could get the best view of the 13,503-foot runway at Burns Flat's former Strategic Air Command base. About two and a half miles long, the wide runway dominated the flat landscape of farms below, as well as the little town beside it. It had been built during the Cold War for heavily laden B–52 bombers ready to scramble at a moment's notice to rain nuclear death on the Soviet Union, half a world away. Seto remarked that he could take off *sideways* on it. Even the parking area beside the air traffic control tower was writ large, covering ninety-six acres in concrete. The place was all but deserted. As we descended, an olive-green C–5 military transport plane touched down on a runway parallel to ours, then immediately lifted off again. Air Force pilots practicing takeoffs and landings constituted just about the only traffic the place saw these days.

If Chuck Lauer and George French and all their engineers and technicians had their way, that would change in the next two or three years as the *Rocketplane XP* took off on regularly scheduled spaceflights. The officials of the Oklahoma Space Industry Development Authority based here at Burns Flat were counting on it. Without a space business to help develop, they wouldn't have much to do, and they had already completed the application process with the FAA to certify the place as a commercial spaceport—an effort that paid off in June 2006 when it became officially known as the Oklahoma Spaceport.

After we landed, Reda Anderson, Misuzu Onuki, and I hooked up with OSIDA director Bill Khourie and operations manager Joe Savage for a tour of OSIDA's nondescript offices, along with a big empty room that OSIDA planned to build out into a mission control center, and a classroom, where a group of elementary school kids were learning about space. The classes were part of OSIDA's mission "to promote space education in Oklahoma." Anderson needed little prodding to step in front of the group of about twenty kids and answer questions put to her by the teacher about her outer space aspirations. "If you go up on the top of this roof here, you see the world from a little bit different view," she told the kids. "If you go up in a small plane, you see it from a little bit different view. If you're in a big airplane, thirty-five thousand feet, you see it from a different view." Her voice took on an excited edge as she said, "What's it going to be like when we go all the way up?" Dramatic pause. "It gives me chills right now thinking about it."

She got chills of a different kind on the flight back to Oklahoma City. I was sitting in the copilot's seat, and Anderson and Onuki were in the back. We were on descending through two thousand feet on approach to the airport when my door popped open.

"The door—" I began.

"The door!" Anderson yelled over the intercom, cutting me off. "Michael, make sure your seatbelt is tight!" Could it be that Reda Anderson, intrepid space traveler, was actually in danger of panicking over a minor in-flight emergency?

I did as Anderson suggested, and I also pulled the door toward me experimentally. It moved easily, though it wouldn't latch. The wind wasn't about it rip it wide open, I decided—in fact it was keeping it mostly closed. I held it in place anyway, for want of anything else to do. I could see the fields and the roads below through the gap between the floor and the bottom of the door. Seto spoke to us in low reassuring tones over the intercom, explaining that there was no danger. Anderson didn't buy it. What else on this airplane was about to pop open, fly off, or otherwise put our lives in danger? She gave Seto hell after we landed. "I hope no one had a camera," Seto said weakly.

Anderson's nerves seemed out of character until I remembered something she had said earlier that day. "When I went down to the *Titanic*," she said, "I did not think that was adventurous at all. You sat down and a guy drove you around, and what's adventurous about that?" A week later, she read in the newspaper about another Russian submarine that had become trapped 600 feet down. Only then, she said, "I started to shake and I got goose bumps." Only then did it hit her just how deep she had been in her own Russian submersible—12,500 feet, in fact, where one mechanical failure or snag would surely have meant her doom. No, she didn't consider herself particularly brave. What she wanted was to experience as much as humanly possible, and taking risks was only a necessary part of that, not a thrilling end in itself.

Which is why she quizzed Rocketplane Kistler chief pilot John "Bone" Herrington extensively about the safety of the ship he would be flying her in.

Herrington, an aerospace engineer, former Navy test pilot, and the first Native American in space, had left the NASA astronaut corps early and taken a pay cut to join Rocketplane Kistler.

He'd flown in space once, on the last space shuttle mission before the space shuttle *Columbia* disintegrated on reentry early in 2003. He knew he'd likely have a long wait before his next ride on the shuttle, and he just couldn't pass up a chance to be part of what he thought would be a historic event—the first suborbital flight to carry passengers into space.

Anderson grilled Herrington about the *Rocketplane XP*'s flight profile, about every aspect of the experience of flying in space, about all the potential dangers. And about whether she'd get a good view out the window. "I'm not wild about getting out of the seat and floating around," Anderson told Herrington. "I'm more interested in the view." There just wasn't time to get up and float around and get used to the sensations of weightlessness *and* take in the sights. Anyway, she could get the weightless experience much more cheaply on a parabolic airplane ride of the sort offered by Peter Diamandis' Zero-G Corp. That suited Herrington just fine; the last thing he wanted to have to do was try to corral floating space newbies and get them strapped back into their seats before reentry. The cramped interior of the *Rocketplane XP* would prevent the passengers from doing much floating, anyway. Anderson actually *liked* the craft's small size and the fact that it could carry a limited number of passengers. That made it more exclusive. Her biggest worry was that in the rush to get to market first and start earning a return on investment capital, old-style NASA "go fever" would make the Rocketplane team cut corners or move too fast at the expense of safety.

But would the *Rocketplane XP* even get off the ground? Dan Erwin, associate professor of aeronautics at the University of Southern California, thought it had an excellent chance. "The performance numbers given by the company are reasonable based on their estimate for vehicle mass at launch," he told me. The team's greatest challenge would be the ship's rocket engine. "There are no show-stoppers in principle," he said of building an all-new rocket engine that would be safe enough to fly on a passenger-carrying vehicle. "But there is an awful lot of elbow grease that has to be laid on the problem." The spaceship's rocket engine was conspicuously absent from my tour of headquarters in Oklahoma City, the spaceport at Burns Flat, and the shop floor in Guthrie. To learn about this most crucial component, the "rocket" in "Rocketplane," I had to head to the Southern California coast.

Polaris Propulsion occupies an unmarked office and warehouse in a development of identical such spaces tucked away among the strawberry fields of Oxnard, about an hour's drive north of Los Angeles. Dave Crisalli, a wisecracking former Navy officer and propulsion engineer for major defense contractor Rocketdyne, works here on rocket projects for heavy hitters like Boeing, Northrop Grumman, and Rocketdyne with his long-time associate George Garboden, who's responsible for machining Crisalli's bright ideas into actual working metal parts. Crisalli and Garboden got their start as rocket-powered entrepreneurs in 1996, when they and other members of the Reaction Research Society (RRS) rocketry club blasted a solid-fueled fourteen-thousand-pound-thrust booster to the very edge of space—about fifty miles, according to their calculations—from the desert north of Las Vegas. That project impressed Boeing enough to hire Crisalli and Garboden as contractors, and Polaris Propulsion was born.

Crisalli's first reaction when George French asked him to build the *Rocketplane XP*'s engine was to turn him down flat. It seemed like just another crazy space dream that wouldn't go anywhere at best, and that might just get someone killed at worst. When French persisted, Crisalli wrote him a thirty-five-page letter detailing the precise terms under which he would work on the project. "The phrase I've used," Crisalli told me when I met with him and Garboden in Oxnard, "and I use it with Rocketplane, is 'there's a difference between thinking out of the box and being out of your mind.' And, you know, it's not that either one is bad; it's just that when you get up any given morning, you gotta know whether this is an 'I'm out of my mind today' day, or whether it's an 'I'm out of my box today' day." After satisfying himself that French and the Rocketplane Kistler engineers in fact had their heads screwed on tight, he enticed a team of Apollo-era propulsion experts back into action to work on the new engine.

The AR–36, as Crisalli was to call it, or Aircraft Rocket–36, would deliver thirty-six thousand pounds of thrust running on liquid oxygen and kerosene. Its regeneratively cooled design (in which kerosene circulates along the combustion chamber's outer wall before flowing inside to be burned) would allow the engine to be fired many times without much maintenance, just like a jet engine. This design had an advantage over rocket engines with ablative coatings that char and flake away to take heat with them; ablatives have to be replaced after every firing of the engines they protect from the intense heat of combustion. When I met Crisalli and Garboden in the spring of 2006, their team had just finished the initial design work on the AR–36, and they were ready to start building it. They planned to test-fire it in the Mojave Desert, the RRS's usual stomping grounds, and they told me they were on track for the *Rocketplane XP*'s inaugural rocket-powered flights in 2008.

Virgin Galactic and Rocketplane Kistler seemed to be racing neck and neck for the honor of flying the first suborbital space passengers. But theirs were by no means the only well-funded ventures in 2006 gearing up for suborbital flights. John Carmack and his Armadillo Aerospace had scaled back their vertical takeoff, vertical-landing X PRIZE rocket ship design to a single-seat vehicle and were still planning to send one of their own into space in the near future. Scaled Composites' neighbor at the Mojave Airport, XCOR Aerospace, was hard at work on a two-seat rocket plane design. This one, like the *Rocketplane XP*, would launch under its own power from a runway but without the encumbrance of jet engines; *Xerus* would be rocket-powered all the way.

And, working in secret, Amazon.com founder and CEO Jeff

...igin's *Goddard* rocket heads back to its Texas hangar after a successful test flight in ...er 2006. *(Courtesy of Blue Origin)*

Bezos had built a self-funded spaceship company of his own called Blue Origin. In 2004, Bezos bought up 165,000 acres of empty ranch land in west Texas, and by 2006 he was flight-testing hardware there for a suborbital tourist spaceship called *New Shepard* (as in Alan B., America's first astronaut, who flew a fifteen-minute suborbital jaunt in 1961). *New Shepard* would take off and land on its tail to send three passengers on automated ten-minute flights out of the atmosphere. Bezos refused to talk to the press about his space venture in any detail, but an environmental impact statement that Blue Origin filed with the Environmental Protection Agency in 2006 revealed plans to launch revenue-producing tourist flights to suborbital space by 2010.

With all this competition, it seemed possible that within the next decade or two, regularly scheduled passenger service to suborbital space could drop to the cost of an ordinarily expensive vacation—a Caribbean cruise, say. Burt Rutan alone proclaimed it his mission "to assure that there'll be hundreds of thousands of people flying outside the atmosphere soon." Not all of those passengers would be space tourists. In fact, most of them wouldn't be. "We think that the future for suborbital is really in point to point, both for people and for fast cargo," said Chuck Lauer. In other words, the biggest market would be for intercontinental travel at rocket speed. Lauer looked forward to the day when passengers could "do an ocean hop in one hour in a vehicle that would be flying at about mach 10." If his prediction came true, it could have as big an impact on travel as the advent of the jetliner.

But while many of the space entrepreneurs worked to make suborbital spaceflight profitable, others dared to aim even higher. One hundred forty miles higher, to be precise: all the way to orbit.

7

ORBIT ON A SHOESTRING

**Elon Musk
and SpaceX**

*Deep Purple's "Smoke on the Water" plays on a
powerful-looking boom box beside me on the scaffold,
which surrounds a seventy-two-thousand-pound-
thrust liquid-fueled rocket engine. With its heavy
metal guitar line and lyrics about a fire in the sky, it's
perfect background music for a rocket engine test
stand. I watch as technicians wearing blue overalls
emblazoned with the white, stylized falcon logo
of SpaceX attach a carbon-and-glass-fiber rocket
nozzle to the six-foot-tall plumber's nightmare that
constitutes the rest of the engine. Turning away
from the engine, I look out at Texas plains land—all
dun-colored earth and tall grasses extending to
the fence perhaps a quarter of a mile away. Ranch
houses sit about a quarter of a mile beyond that,
and beyond that, just more empty grassland.*

It's mid-January and the air is frosty, making my fingers and toes ache. At least I'm out of the wind here. Tom Mueller, SpaceX chief of propulsion, joins me at the rail and points to a brick house 45 degrees off to our right. "The people in that house love us," he says with a chuckle. That's where the noise of the downward-facing rocket engine sounds the loudest, not in a straight line from the exhaust deflected off the concrete pad below us. "Yeah," agrees technician Kenny Thomas, taking a break from work on the engine to join us. Just the sound of the engine here when it's firing, Kenny jokes, could kill you. "Or if it didn't kill you, you'd wish you were dead." Even way out in the town of McGregor, Texas, miles away, the rocket blast rattles windows. "You woke my baby last night," the police officer on duty told a SpaceX engineer who called the police station to warn her of an impending test-firing.

"Well, I'm really sorry about that, ma'am," said the engineer, an unfailingly polite native Texan.

"And you're going to wake my baby again tonight, aren't you?"

"Well, I suppose so, ma'am."

The deflected rocket blast "mows the grass," as Mueller puts it, all the way out to the fence line. I can see that the concrete is deeply scored in a straight line out to where it ends and the grass begins a couple of hundred feet away. And beyond that, there's plenty of bare dead earth, with the line continuing as a shallow depression.

Later, a turkey buzzard circles far above the test site as a liquid oxygen tank fills with nonflammable liquid nitrogen for a structural test. The SpaceX engineers have noticed that the buzzards tend to show up as super-cold liquid gases hit the comparatively warm metal of their tanks and steam bleeds off from pressure relief valves with a shrill scream and a thin cloud of condensing vapor. The buzzards have learned to read the signs of an impending rocket engine test and its accompanying jet of flame, lethal to any small animals caught in its path.

The engine is Mueller's baby, financed by SpaceX founder Elon Musk, a thirty-three-year-old physics major from South Africa who cleaned up during the computer tech boom of the 1990s and is now using his personal fortune to build (he hopes) cheap access to outer space. It's called the Merlin, and it will power SpaceX's first rocket, the Falcon 1—as in the *Millennium Falcon* of the *Star Wars* movies. Falcon 1 will be considerably less advanced than the fictional starship, which can transverse interstellar distances at super-light speeds, but it will be impressive nevertheless.

After launch from Vandenberg Air Force Base near the Central California coast, Falcon 1 will hit nine times the speed of sound before the single Merlin engine in its first stage shuts down at 380,000 feet in altitude (seventy-two miles). The first stage will then separate from the rest of the ship and reenter the atmosphere for a splashdown in the Pacific Ocean off the coast of Baja California, while the second stage, powered by a smaller engine called the Kestrel, will push the rocket's payload, a small (fifteen-hundred-pound) satellite, the rest of the way into orbit.

Traveling three times faster than *SpaceShipOne* on engine cutoff and reentry, Falcon 1's first stage, made of aluminum rather than *SpaceShipOne*'s carbon fiber composites, will be subjected to correspondingly greater stresses, though without any wings or other control surfaces to risk shearing off or pesky pilots to object to severe g loads. Amazingly, the first stage—along with, of course, the Merlin—is designed for repeated use. It will land in the ocean on parachutes, there to bob up and down getting good and soaked with corrosive sea water for the two days it will take a recovery ship to get it back to harbor. After harnessing the power of exploding kerosene fed by liquid oxygen, the Merlin will endure the cold plunge into the ocean while still hot. That's a lot of abuse for a carefully assembled collection of precision machined parts that all have to function in perfect equilibrium to avoid a "rapid unplanned disassembly," as rocket engineers euphemistically term a catastrophic rocket engine explosion. But SpaceX's engineers are confident that Merlin's valves, pipes, turbo pump, and other parts, which are made of high-quality nickel alloy and other corrosion-resistant metals, could withstand not just a couple of days in the ocean but several *months* without suffering significant damage.

This impressive feat is one of the keys to Musk's plan to undercut every other launch provider in the market. The least expensive of Musk's competitors, Orbital Sciences Corporation, charges somewhere in the neighborhood of $30 million for its air-launched Pegasus rocket. Pegasus is the kind of small satellite launcher Musk envisions will make his company independent of his $100-million-and-counting investment. Launching a big satellite—for instance, the type used by TV networks for broadcasts from high orbit—costs far more: on the order of $50 to $75 million using American launchers like Boeing's Delta II. Musk plans to charge a mere $6.7 million per launch.

That might sound overly optimistic until one considers that every other satellite launcher ever built has no reusable parts, which contributes greatly to their huge price tags. These extremely expensive machines get thrown away like so many discarded rifle bullet casings.

Imagine throwing away a Boeing 737 jetliner after just one passenger-carrying flight across the Atlantic, and you'll see part of the problem facing the satellite launch industry. To recoup their costs, let alone make a profit, airlines would have to charge each passenger not only for the cost of the trip but also for the cost of building the airplane. Tickets could top $400,000 each, which would mean far fewer passengers and consequently far fewer airplanes flying across the Atlantic. So it is with satellite launchers, which sharply limits their numbers and hence the services they can provide. But what would happen if the price to launch a satellite were significantly lower? What if even a midsize company or a university could launch a satellite? No one knows what new markets for satellite launchers will open up, but Musk is confident he'll get more than enough business to fund his real mission.

"The reason I started SpaceX," he told me, "was for human spaceflight, with a very long-term objective of enabling human colonization of Mars." Why not start with a less ambitious destination like the moon? "The moon is not really a good place to have a second human civilization. The moon is analogous to the Arctic. If Europe is the Earth, then the moon is the Arctic and Mars is North America. The moon is very barren. It's very resource-poor. It has no atmosphere at all. It has very low gravity. It has a day which is twenty-eight days long. So there are just a lot of things about the moon that make it a very difficult place to create another human civilization."

As a stepping stone on the way to getting to Mars, a bigger version of Falcon 1, called the Falcon 9 and powered by no less than *nine* Merlin engines in its first stage, would loft not just satellites into orbit but five to seven astronauts at a time. And then, says Musk, the markets would *really* open up. "Transportation is such a fundamental input to any business plan that I think there are going to be a lot of applications that people can't even guess at today." But if he had to hazard a guess, "the most significant business is probably people that want to go to space. You know, personal space travel."

Musk was born in 1971 in South Africa to a South African father and a Canadian mother. His mother was a model and nutritionist; his father, an engineer. His father helped inspire in Musk a love for high technology—although as Musk later recalled, he "was a luddite with respect to computers and refused to use them in his work." Musk saved up his allowance to buy his first computer and taught himself to program it by the age of ten. When he was twelve, he sold his first commercial software, a space battle game called Blastar. He left South Africa for Kingston, Ontario, in Canada at seventeen, in part to avoid compulsory military service. "I don't have an issue with serving in the military per se," he said later, "but serving in the South African army suppressing black people just didn't seem like a really good way to spend time." Still, he probably would have left anyway, because he ultimately wanted to live in the United States. "Particularly if you're interested in advanced technology, the best place in the world is the U.S."

In Canada, Musk attended Queen's University, scraping by on as little as $1 a day until he landed a scholarship to attend the prestigious Wharton School of Business at the University of Pennsylvania. In college he thought hard about when he should do with his life after graduation. He came up with three areas he wanted to get into because, as he said later, "they were very important problems. One was the Internet. One was clean energy, and the third was space." After graduating with a bachelor's degree in economics, he stayed on another year and picked up a bachelor's in physics as well—seemingly a perfect foundation for a future as a space entrepreneur. Graduate school was Stanford University, where he also studied physics—for all of two days.

Stanford was at the epicenter of the computer technology boom that was just hitting its stride when Musk moved there in 1995. He looked around, saw opportunity everywhere, and decided he couldn't let it pass him by. He asked to be allowed to defer school for just a quarter, in order to start an Internet company. He figured that like most new businesses, his would fail to get off the ground, and he'd be able to go back to school after his little experiment in applied economics. His graduate advisor knew better, and in his first and last conversation with Musk, he told him he thought he'd never come back.

He was right. After humble beginnings during which Musk had nowhere to sleep but his office and he had to shower at the local YMCA, the startup he founded in 1995 took off like, well, a rocket. The company, Zip2, enabled newspapers to combine editorial content such as movie and restaurant reviews with consumer business directories. In 1999, Musk sold the company to the Compaq computer maker, which added it to its AltaVista Web search engine. Twenty-two million dollars richer, Musk jumped into another venture that became the PayPal online payment service. Musk was only thirty years old in 2002 when he cashed that one in for a breathtaking $1.5 billion in eBay stock. Musk was PayPal's largest shareholder, and his personal stake in the sale (to online auction service eBay) came to over $200 million. One estimate had Musk's net worth then topping out at over $300 million.

Flush from his Internet success, Musk turned his attention to one of the other items on his list of important things to do: space. His first idea was to build a robotic Mars lander he called Mars Oasis. This was to be a kind of Martian terrarium that would germinate seedlings from Earth in a nutrient gel and, he hoped, produce the first plants to be grown on another world, albeit in an enclosed environment. He thought the project could produce useful scientific data while also sparking increased public interest in space. The estimated $20 million for the Mars lander did little to deter Musk. But the cost to launch it on a Delta IV blew him away: two and a half times as much as the already fabulously expensive (but to him, affordable) project itself. He checked out the possibility of launching his experiment on a cheaper Russian rocket, but the labyrinthine and at times shady politics he would have had to negotiate put him off of that possibility.

So it was that Musk encountered the single biggest obstacle to working in space—the cost of getting there. And an intriguing possibility occurred to him: Why not build his own launcher? Even better, why not make that his business so that other like-minded people could launch their own experiments? Just how feasible would it be to build a cheaper launch vehicle, anyway? To find out, he hired the firmer chief technical officer of Orbital Sciences, Mike Griffin (who later became head of NASA), to put together a formal study of the problem. What he learned from the study was that there was no reason on Earth why orbital launch vehicles had to be so expensive. "It's just that those who have built and operated them in the past have done so with horrendously poor efficiency." The problem of relatively affordable access to space, in other words, wasn't technological; it was organizational. The big aerospace companies had a lock on the launch market, and without competition they had no incentive to charge anything less than absolute top dollar for launch satellites. They were hugely inefficient as a consequence, driving costs even higher. Musk figured that a small nimble start-up company could grab not only a share of the existing launch market but also a market that previously couldn't afford launch services.

Not that making yet more money for its own sake was Musk's goal. "If we ultimately wish not to go in the direction of the dinosaurs," he told me, "we will have to become a multiplanetary species." In other words, as long as we are confined to a single planet, we run the risk of being wiped out by a planet-wide catastrophe, like the asteroid strike that doomed the dinosaurs. "And the dinosaurs did not face the risk of self-annihilation that we do; all they could do is try to eat each other one at a time." For Musk, settling the solar system was nothing less than a species imperative.

Musk already had the startup capital for a space launch business. Now he needed the technical expertise. So he went shopping at the country's oldest rocketry club, the Reaction Research Society (RRS), based in the Los Angeles area, with ongoing rocket launches in the Mojave Desert. This was the same club through which Rocketplane propulsion contractor Dave Crisalli had launched his homemade solid-fueled rocket to the edge of space. While the RRS was strictly an amateur affair, its members included rocket engineers from major aerospace companies working on missiles for the Department of Defense. Musk knew it was a class act, full of qualified engineers willing to try something new. Sure enough, there he met Tom Mueller, head of propulsion for defense contractor TRW and one of Crisalli's cohorts in his 1996 edge-of-space rocket launch. Mueller just happened to have in his garage the world's largest homemade liquid-fueled rocket engine. "Can you build something bigger?" Musk asked him.

Together, Musk, Mueller, and another rocketeer, Chris Thompson, worked out how they would beat the big aerospace companies at their own game by building cheap rockets and launching satellites at cut-rate prices. They would set up shop in El Segundo, California, near the Los Angeles International Airport, right in the shadows of Boeing, Raytheon, and the other big defense contractors that had dominated the satellite launch industry from the beginning. And they would poach the best young ambitious engineers and managers right out from under the big guys, who would never know what had hit them. Especially when SpaceX started launching astronauts along with satellites.

Wearing the simple dress shirt and khakis that constituted something of a uniform in the software industry, Musk when I met him in early 2005 could easily be mistaken for one of his employees. Even his office was just a doorless cubicle like dozens of others at SpaceX headquarters, albeit somewhat larger than those of his employees. The offices occupied the second floor of the SpaceX Main Building, as it was known—a white concrete two-story building across the street from a condo complex and in view of the control tower at LAX. Musk shared the second floor with the structural engineers, who were responsible for designing the aluminum tanks and outer shells of the rockets. A framed photo of Muhammad Ali having just knocked out Sonny Liston hung over Musk's computer workstation. It gave the only hint of the sheer audacity of the soft-spoken, unassuming young man who worked in this cube. The photo helped him focus on his mission to steal some of the thunder from the aerospace giants headquartered all around him.

On the main floor were the cubicles of the propulsion engineers—the rocket scientists. The back half of the building was taken up with a machine shop open to both stories. An American flag hanging high up on the shop's front wall overlooked the gleaming seventy—foot aluminum tube of a Falcon 1 rocket laid out on its side undergoing preparations for its journey to the launch pad at Vandenberg Air Force Base, one hundred fifty miles up the California coast. Engineers and technicians worked on the shop floor in black SpaceX shirts with stylized flames on the long sleeves. There were no windows, but steel loading dock doors had been rolled open to let in the bright Southern California sunshine.

Musk invited me to peer into the guts of his rocket while he filled me in on his business model, which he insisted wouldn't need any exotic technologies to work. "You know, Ford didn't invent the internal combustion engine," he said. "But he found out how to make one at low cost, and that's the appropriate analogy here. We didn't invent the rocket engine; what we're trying to do is figure out how to make it low cost." How would Musk pull that off? There's no single solution, he told me. "Whenever some company tells you that XYZ magic silver bullet is the reason that they're better or cheaper," he said, "they're not telling the whole truth; they're just using that as a simplification."

Musk had identified five major drivers of launch vehicle costs and had set about bringing each one of those down. The first, and perhaps the biggest, was overhead. Looking around the Main Building, I could see for myself that Musk ran a tight ship. He had fewer than a hundred employees, including all the engineers, machinists, and associated support staff like the receptionist, administrative assistants, and a public relations officer. "We are an extremely low-overhead company," said Musk. "If we simply handed our blueprints to a Boeing or Lockheed, I think the price would at least double, if not more."

There was the rocket itself, with three major components contributing to its cost: engines, structures (the actual body of the rocket as well as fuel and oxidizer tanks), and avionics—the sophisticated computers and software that controlled the other components to guide the rocket through the air to space and into orbit.

And then there was the launch operation. Musk laughed as he told me how a Lockheed Martin representative boasted to him of his company's "lean" launch crew: only three hundred people. "Now what are those people doing? I can't tell you." Falcon 1's launch crew? Twelve to fifteen people sitting in a custom-made trailer at the launch site.

Paring down the rocket itself was Musk's surest way of reducing its three contributing cost drivers (engines, structures, and avionics). To begin with, he and Mueller and the other engineers conceived the simplest possible design that could still reach orbit: just two stages with no strap-on boosters. Simple, compared, for instance, with the Delta II's two liquid rocket stages plus a solid upper stage and assorted strap-on boosters. "Each stage is to be really thought of as its own spacecraft," Musk told me. "So even doing something as innocuous as going to three stages, or two stages plus strap-on stages, is a huge percentage increase in complexity and cost." The Pegasus was arguably even more complex, with two solid-fuel stages and a carrier plane like *SpaceShipOne*'s that had to be maintained even between launches. And since Pegasus got dropped off horizontally by its carrier plane, again like *SpaceShipOne*, it needed wings and airplane-style control surfaces to orient itself. More complexity, and hence higher cost, according to Musk.

After settling on as simple an overall design as possible, Musk and his engineers focused on streamlining the engine. They started with a standard liquid-fueled turbopump design.

Liquid-fueled rocket engines must burn fuel at a ferocious rate to accelerate their payloads to the 17,500 miles per hour needed to reach orbit, and both the fuel and the oxidizer must flow into the combustion chamber fast enough to keep up—in other words, at high pressure. One obvious solution is to keep the fuel and oxidizer under high pressure in their respective tanks. But that requires extremely strong—and heavy—tanks. Extra weight is the mortal enemy of spaceships, since every pound requires many more times that amount in fuel to reach orbit. Therefore, many liquid-fueled rocket engines, including the design Mueller and his engineers decided on, use a high-speed pump called a turbopump to shoot high-pressure fuel and oxidizer into the combustion chamber. Merlin's kerosene and liquid oxygen are lightly pressurized in their tanks with nonflammable helium from separate onboard tanks to get them flowing into the turbopump. To keep things as simple as possible, Mueller and his engineers decided on a pintle injector design, which uses a single fuel injector to mix the kerosene and oxidizer coming out of the turbopump, instead of the many injector holes used on other rocket engines.

A gas generator burns a relatively small amount of fuel and lox to generate hot gas, which then blows past a turbine wheel in the turbopump to spin it up to the high speed it needs to pressurize the kerosene and liquid oxygen. Exhaust from the combustion chamber rushing out the rocket's nozzle propels the rocket skyward, but the gas generator produces its own exhaust.

With six hundred pounds of thrust, Merlin's gas generator exerts nothing like the seventy-two thousand pounds of thrust put out by the combustion chamber, but it's still enough to help steer the rocket in flight. The exhaust flows through a small moveable rocket nozzle hanging off the side of the main engine to give the ship roll control, serving the same function as an airplane's ailerons. Merlin's main rocket nozzle itself moves on a gimbal for thrust vector control, or TVC, to control the rocket's other two axes of motion, pitch and yaw (on an aircraft, controlled by the elevators and rudder).

Mueller and crew further simplified the engine design by tapping into the turbopump for high-pressure rocket fuel to use as hydraulic fluid to power the TVC gimbal. As in other areas of the Falcon 1 design, this feature not only makes the ship less complex but increases reliability as well; the rocket can't run low on hydraulic fluid as long as the engine has propellant to burn.

In the Avionics Building, I met aerospace engineer Brian Bjelde, a slightly rotund twenty-four-year-old wearing a pink Hawaiian shirt replete with postcard images of sailboats, palm trees, and bikini-wearing girls on towels. He showed me the brains of the Falcon 1, the avionics tray that would go in the rocket's upper stage. It was a metal hoop, as big around as the rocket itself (five and a half feet), that would fit just below the payload. It now rested on a stand about waist high, where Bjelde and his colleagues could easily get at the computers, relays, and circuit boards bolted to its inner surface.

Bjelde had been at Jet Propulsion Laboratory (JPL), which builds NASA's space probes, and was working on his master's in astronautics when he got poached by SpaceX. At JPL, Bjelde worked in a clean room in a head-to-foot suit where "you have to identify people by their eyes," he told me. "It was really neat and really prestigious, but way too much bureaucracy. It's ridiculous, the hoops you have to jump through." He was working on vanishingly small micronozzles, tiny rocket nozzles just one fifth the width of a human hair, and peering through a microscope on specialized alignment equipment to even see his workspace. One day the alignment machine broke down. "It's a half-million-dollar piece of hardware," Bjelde said, and the machine had to be sent back to its manufacturer in Germany for repairs. "All the paperwork and documentation, all the bureaucracy—I'm out of a job for six months. I sit around and do nothing waiting for the system." Frustrating, to say the least.

Here at SpaceX, all Bjelde had to do was call up Musk if he needed a replacement part. "I say, 'Elon, I need something,'" and "*Boom*, you got it." Musk not only personally controlled the resources Bjelde needed but also trusted him to know what he needed and when he needed it without making him fill out a lot of paperwork. "And that's a nice thing, too," Bjelde told me, "getting trust at my age; you don't get that in the big companies." SpaceX was paying for the completion of his master's degree, but Bjelde didn't know whether he ever would make it back to school. He was too busy to devote any time to his course work—which, actually, he didn't mind at all. "I mean, I get to get my hands dirty. I've learned so much in such a short period of time. It's nothing less than awesome."

One of Bjelde's responsibilities was the so-called flight termination system, essentially a fancy garage door opener with associated antenna, receiver, and electronics that cost $100,000. The military personnel at Vandenberg Air Force Base, from which the first Falcon 1 was scheduled to launch, required not just one but two of these (in case one failed). Among the SpaceX crew in the launch trailer would be an Air Force man acting as range safety officer, his hand poised over a switch to kill the rocket if it veered off course and headed for, say, Los Angeles. Most big rockets, including the space shuttle, have such an onboard system. In the shuttle's case, the system includes shaped explosive charges on the external fuel tank and on both solid fuel strap-on boosters. If the range safety officer determines that an out-of-control rocket could hit a populated area or offshore shipping, he'll send first an "arm" then a "fire" command to activate the system and destroy the rocket.

The shuttle's system was used during the 1986 *Challenger* disaster when one of the strap-on boosters sprang a leak of hot gases and the resulting side-angle exhaust burned through the giant external fuel tank and touched off an explosion. The solid-fuel boosters took off on their own at that point, still intact, and the range safety officer back at Cape Canaveral blew them to bits before they could hit the ground or the ocean. It's just one of the many worries that plague space shuttle astronauts as they strap in for a launch: that if an in-flight accident doesn't kill them on the way up, some Air Force officer back on the ground might just decide to finish the job. Only an indicator light on the commander and pilot's control panels would warn the crew of impending doom when the range safety officer sent the "arm" command. The astronauts would have no chance of survival if the range safety officer hit his "fire" switch. The Falcon 9, on the other hand, will have an escape tower on top of the rocket that will pull the space capsule with the crew inside away from the rest of the rocket in an emergency. This is the same type of escape system that flew on American spaceships before the shuttle and that still flies on the Russian Soyuz.

The flight termination system Bjelde engineered for Falcon 1 didn't include explosives. Because the rocket was relatively small, with a low potential for widespread mayhem if it went off course, Air Force officials approved a nonexplosive "thrust termination" system. "So instead of exploding the vehicle," Bjelde explained, "we command a couple relays to open, and they remove power to the valves." The valves feeding propellants to the rocket engine always remain closed unless pneumatic pressure opens them. With power gone, the valves snap closed, the flow of fuel and oxidizer to the engine shuts off, and the engine shuts down. At that point, said Bjelde, "the thing drops into the ocean" off the California coast. "It's always expelling propellant, so if it's close to the end of the first stage, you're going to have a big empty tank and a full second stage."

While SpaceX's engineering and fabrication work went on in El Segundo, the task of flight-qualifying the rockets (testing, testing, and more testing) went on at an old Navy rocket test site outside of the little town of McGregor, Texas. Making the two-and-a-half-hour drive from Dallas on Interstate 35, I felt as if I had landed in another country. The feeling wasn't entirely unwarranted, according to a bumper sticker I saw on a pickup truck that out-massed my little economy rental car by several times. "I'm from Texas," it said. "What country are you from?" A sticker on an equally massive SUV read "God Bless John Wayne." I felt like a midget in a land of giants. Everything about Texas seemed larger than life, including the Bible shop I passed on Highway 84 in Waco called Compass, the "Superstore for the Christian Lifestyle."

On a country road outside the town of McGregor, I drove past an empty guard post and a faded sign put there by the U.S. Navy to warn off interlopers. Musk had chosen the area for the same reason the Navy had: the miles of empty land around it, the buffer needed by any rocket test operation to avoid hitting innocent bystanders with exploding rocket motors and out-of-control missiles.

Joe Allen, one of SpaceX's machinists working here, knew

more about errant rockets than most. The easygoing fifty-year-old African American had worked at the test site for twenty-six years, first in 1978 for a company called Hercules, for which he built solid rocket motors under contract for the Navy. Once a Sidewinder air-to-air missile tore loose from its moorings on a test stand, went sailing overhead, nosed over and buried itself two feet in the ground, where it sat sputtering and fussing like the angry snake it was named for. Allen laughed as he described the shock of the rocket scientists, who had thought that they had rendered the missile flightless by removing its wings.

Allen took everything he saw here in stride, and the younger SpaceXers respected him as a wise old-timer. One of them, a freshly minted structural engineer named Florence Li, the only woman working at the site, believed that Allen regularly killed rattlesnakes with his bare hands. Joe himself told me in his slow Texas drawl that he had killed eighteen rattlesnakes out here the year before. "How'd you do it?" I asked him.

"Well, I have a shotgun, but I never get a chance to use it. I use a shovel."

"They've been telling stories about you. They say you kill rattlesnakes with your bare hands."

He chuckled. "No. But I've been close enough to."

After the Navy left the site, a startup called Beal Aerospace took it over. The company was headed by Andrew Beal, like Elon Musk a self-made man with plenty of money. The company went bust in 2000 after burning through $10 million a month out here, according to a rumor that Tom Mueller had heard. After Beal saw it was going to cost him $300 million to develop a rocket capable of sending sizeable satellites into orbit, he called it quits, leaving behind a lot of useful infrastructure for SpaceX, including a three-story concrete tripod with legs as big around as redwood tree trunks. Nicknamed the BFTS (for "Big Fucking Test Stand"), it would be used to test Falcon 9 and its nine integrated Merlin engines.

After Beal left and there was a break in activity at the test site, Allen collected a degree in computer programming, but he went back to working on rockets when SpaceX moved in.

SpaceX's control center was in a concrete bunker partly buried in earthworks as protection against rapid unplanned disassemblies. It had air conditioning and heating units on top and satellite TV dishes on the side. The facility's mascot, a lovable black mutt named Rockette, came running up to greet me with her tail wagging as I pulled into the parking lot near the bunker's entrance.

Past a steel blast door, through a small machine shop where engineers and technicians worked on rocket nozzles, on the other side of an inner blast door looking like something that belonged on a nuclear submarine, six forty-two-inch flat screen monitors covered the front wall of the control room. Three rows of long tables held computer workstations with their own flat screen monitors. The propulsion engineers sat up front, near the big monitors, which showed camera views of the test stands, while the structural engineers sat in the back rows. I set up my laptop computer at an empty spot among the structural engineering stations and was able to connect to the Internet as soon as I booted up. Except for the lack of windows and the concrete walls, the place could have been any Internet startup from the heady days of the dot-com boom, complete with a kitchen area well stocked with munchies and an expensive-looking electric massage chair in one corner.

Ordinary computer network cables linked the bunker to the test stand holding the Merlin engine to be test-fired that day, as well as to fuel and lox tanks undergoing structural tests. To test the strength of the tanks, structural engineers pressurized them with rocket fuel and nonflammable liquid nitrogen in place of lox. The test went well until the fuel tank popped a weld and began weeping RP–1 from the bottom, soaking the concrete the tank stood on.

After draining the remaining fuel, Allen, Li, and the rest of the structures crew removed the bottom of the tank so they could climb up a ladder and peer inside with flashlights, looking for the bad weld. As I stood hunching my shoulders against the chill below, they waved me up onto the ladder so I could poke my own head inside and sniff the gasoline reek of the rocket fuel.

Back in the control room, chief of testing and launch operations Tim Buzza jiggled in the massage chair as he told me that the structural test actually went quite well. "We took it to MEOP [Maximum Expected Operating Pressure] times 1.25 without it bursting. That's good." Josh Jung, a recent engineering school grad, was particularly excited about the test. "They had me run the structural tests for the first time today," he said. At a big aerospace company, he would have had to work his way up through the ranks and pay some serious dues to get an assignment like that.

The day's main event was to be Merlin's first full-duration burn. The rocket engine needed to run for 162 to 166 seconds to push the Falcon 1 high enough for the Kestrel engine in the second stage to take the payload the rest of the way to orbit—what the engineers called a mission duty cycle. The engine had never run that long before. Though three previous attempts had been made, technical glitches had always halted the runs before the fuel could run out.

After tinkering with the engine for most of the day, and then pressurizing the test stand's lox and fuel tanks, the SpaceX crew was ready to give Merlin another shot. All sixteen crew members—both the structure and the propulsion teams—went into the bunker and pulled the blast doors shut when the lox tanks started to fill; safety rules required it. Rockette stood in the bunker's machine shop, tail wagging forlornly as the inner door swung to; she wasn't allowed in the control room.

Test engineer David Yarborough took his seat at his workstation at the front of the control room, just below the big flat screen monitors. He would use his computer to control everything on the test stand via the computer network—to open and close valves on the engine and tanks, for instance—and to give the countdown and then enter the commands to fire the engine. Test conductor Kent Harris settled in to Yarborough's right. Reading from a checklist, he would issue the orders Yarborough would follow to prep and fire the engine. Test director Jeremy Hollman, responsible for pretest planning and data collection during the test, sat to Harris's right, ready with a gray box with a red abort button on top. Called the soft abort, that switch was the last resort if the ostensibly tame rocket engine became a rampaging beast and automatic aborts failed to shut it down. Merlin had to burn thirty-four hundred gallons of liquid oxygen and twenty-nine hundred gallons of kerosene for a full-duration burn. All that fuel and lox could make for a spectacular and highly destructive fireball if allowed to explode all at once. As it was, a water tank would have to dump six thousand gallons per minute onto the concrete pad below the test stand to protect it from destruction by the rocket exhaust when the engine fired.

The engineers waited for the liquid oxygen to chill the engine down to its ideal operating temperature before "We're ready to rip, man," said Mueller. "Let's run it."

The engine test engineers, all young men, went through their checklists to make sure the engine's valves and actuators were in their correct positions for the test run. "TE [test engineer], confirm fuel pre is open," Harris called out.

"Fuel pre is open," responded Yarborough.

"Copy that. All right, also the TVCI."

"TVCI shows open."

"All right, TE, let's verify roll control nozzle."

With Harris reading off his checklist and Yarborough responding by checking his workstation display for indicators of all the myriad settings on the engine outside, the little drama smoothly played through its script until the thrust vector control hydraulic return valve stubbornly refused to indicate open. "That valve is not open," confirmed Harris.

Someone softly swore.

"I remember calling it out," said Hollman, not entirely helpfully.

All that could be heard in the room at that point was the humming of the computers as the propulsion engineers considered that the entire day's work could be wasted.

Now the managers stepped in. "Which valve is not open?" asked testing chief Buzza.

"TVC hydraulic return valve," replied propulsion chief Mueller.

"We have no ability to command the TVC with any accuracy," explained Harris.

"So we gotta dump," said Mueller. The valve was hand turned, requiring someone on the test stand to open it. But safety rules required that the liquid oxygen tank be dumped before anyone left the shielded bunker. Dumped. As in vented into the atmosphere. All thirty-four hundred gallons of it, vaporizing into thin air as soon as it was released from the test stand's tank; it couldn't be drawn back into the heavy supply tank to which it had been delivered earlier that day. So even if the engineers did open the valve, they couldn't run the engine. It was now after 6:00 on Friday night, and there'd be no more lox deliveries today.

"I mean, the igniter will still fire, no problem," said Harris, thinking out loud. "The roll nozzle, we don't care if it still cycles, so that's not a big deal. But the TVC has no idea where it's at."

"No, you can't bang it back and forth even if it stops; it won't hold itself steady," agreed Mueller.

"It's one return valve?" asked Buzza.

"It's the one right on the bottom of the fuel tank," said Hollman.

"And we closed that when we were out there?"

"I closed it when we were pulling the—something."

Everyone stayed cool. They kept their voices low, working the problem. But there just didn't seem to be any way out. Until someone dared to suggest it: someone could run out there and open the thing without dumping the lox, safety rules be damned, and the test could go on as planned.

Letting one of his engineers take that risk was out of the question for Buzza. If anyone was going out there, he would be the one. Mueller dismissed the idea immediately, but Buzza insisted. And after some quietly heated discussion, some of it out of earshot of the others in the machine shop, Buzza and Kenny Thomas took off.

On the big monitors at the front of the control room, we could seem them climbing the scaffold of the test stand, moving back and forth in front of the cameras. They worked fast in the no-man's-land around the fully loaded rocket engine while everyone in the control room watched, hardly daring to breathe.

Then, "Okay, we're falling back now," Kenny said over his walkie-talkie. They were back only a couple of minutes after they left, though it seemed much longer. They sealed the blast doors behind them, and the test prep resumed as if nothing had happened.

"Temps look good. We're ready," said Hollman.

"All right," said Harris. "Lox fill is closed."

"Want me to arm soft abort?" asked Hollman.

"Not yet," said Harris. "TE, open lox pre."

"Lox pre is open," said Yarborough.

"All right. TD, verify you're ready to run test and arm soft abort."

"Okay, temps look good," said Hollman. "Soft abort is armed."

"Copy that. TE, close lox vent."

"Lox vent is closed," said Yarborough.

"Copy that."

"TE, arm pad control auto sequence."

"Okay, auto sequence is armed."

"Copy that. Waiting for water."

Silence. Then, on the monitors we saw the water pour down. Over the audio monitor it sounded like a low-fi waterfall.

"All right, there it is," said Harris. "TE, arm mission computer auto sequence."

"Mission computer is armed."

"Copy that. Fire mission computer auto sequence."

"Started."

"All right. Give us the countdown."

"Minus fifty. Minus forty-five. Minus forty. Minus thirty-five. Minus thirty. Minus twenty-five. Minus twenty. Minus fifteen. Minus ten. Nine. Eight. Seven. Six. Five. Four. Three. Two. One. Zero."

The engine coughed and belched out a great gout of orange flame. It roared like a caged wild animal over the speakers, and a low powerful rumble penetrated the earth and concrete walls of the bunker. It sounded and felt like a small earthquake.

For all of five seconds.

The engine quit with a parting lick of flames around its nozzle, and then silent white smoke drifted across the plains in the camera views.

"Aborted," called out Harris.

"Aborted at 5.37 [seconds]," reported Yarborough. "TTI high," he said, reading off the abort code from his monitor. The computer, sensing a problem, had shut down the engine automatically. Now it was up to the crew to figure out just what that problem was. First, though, they had to secure the engine, close the fuel and lox valves, and turn off the cascade of water still pouring over the flame trench.

Afterwards Buzza and Mueller conferred with their engineers. The fault had been with a thermocouple that measured the temperature of the hot gas driving the turbine—a little sensor called TTI (for Turbine Inlet Temperature). It had failed, and when it did, it gave a false reading of critically high temperature, which in turn tripped the abort from the computer. Hence, "TTI high" as the abort code. Mueller sighed as he looked over the data. "Man. Perfect run except for a goddamn $30 thermocouple. Shit."

After some discussion, the engineers decided that a sensor that measured the temperature of the hot gas blowing *out of* the turbopump could compensate for the failed sensor measuring the turbopump's inlet gas. Top off the lox tank from the storage tank, Buzza told Hollman, and, most importantly, "Take TTI out of the abort."

By 6:30 the crew was ready for their last chance at a full-duration burn. The three test engineers again took their places at the front of the control room. "Open TVC press," said Harris.

"It's open," confirmed Yarborough.

"Verify roll control."

"It works."

"Water mains on."

"Water mains open. I'll do it at T minus 20."

Musk chose that moment to call in from his private jet to check on things. "Talked to Elon," Buzza reported to the crew after he spoke with him. "He said take all the aborts out." Everyone laughed, breaking some of the pretest tension.

After the engineers had run through their checklists, Yarborough again counted down.

As before, the nozzle erupted a brilliant font of orange sunfire. The floor trembled as with a distant earthquake.

"Five seconds," said Harris.

"Domes are cold," reported Hollman. "Chambers look good."

"Ten seconds."

"TGI's sitting at one thousand," said Hollman.

"Fifteen seconds. Twenty seconds. Thirty seconds."

"Okay, we got one dome coming up through seven hundred," said Hollman, reporting a potentially show-stopping rise in temperature in some of the engine's hardware.

"Forty seconds," said Harris.

"Slowing down," said Hollman.

"Keep an eye on it," said Mueller.

"It's holding at seven-fifty."

"Fifty seconds," said Harris.

Something bright white began to dance into view on the test stand scaffolding. "Okay, we got something rattling across the deck out there," warned Buzza.

"It's ice," said Harris.

"No, it's something big," said Buzza. "Is there trash out there?"

"It's ice," insisted Harris.

A moment later, more of the mysterious substance shuffled into view, vibrating across the scaffolding, along with a plastic kitchen trash container; someone *had* left garbage out there. No matter. It wouldn't hurt anything. The engineers relaxed slightly. "Seventy seconds," said Harris.

"The domes are coming down in temp," said Hollman.

"Eighty seconds."

Still, the engine roared, as far as I could tell watching it on the monitors, as unstoppable as a volcanic eruption.

Josh Jung, manning a set of electric breaker switches on the wall to the right of the engineers monitoring the engine, breathed a barely audible "Oh man."

"One-twenty," Harris called out. "One-thirty."

"DPFT is down to point eight," called Hollman, referring his readout for Delta Pressure, Fuel Tank—the rocket engine's gas gauge. It was running low, which meant a full-duration burn was almost in reach.

"One-forty," said Harris. "One-fifty."

"DPFT is point four," said Hollman. Then, "There's a grate loose," he said as the engine's fierce vibrations shook loose a piece of the scaffolding's flooring from its welds and sent it dancing with the trash.

"It's all right, let it go."

"One-sixty."

At 162.2 seconds, the engine shut down with a pop of electric circuit breakers tripping. Sudden silence.

"We aborted on DPFT." said Hollman. The engine had run through its gas tank, mission accomplished.

"Copy *that*!" Exulted Mueller.

Clapping. Delighted laughter. Handshakes and backslaps.

There was no reason to hold on to the liquid oxygen any longer, so "Do lox purge," said Mueller. "Full duration burn, man, right there! Lox purge!"

"Hit lox purge," said Harris. "Disarm soft abort."

"Already disarmed," said Hollman, still laughing.

"Take that, mother fucker!" cried Mueller, prompting more delighted laughter. "Call Elon, tell him we just ran a full duration!"

"DTS–1 closed," said Harris, running through his checklist for putting the engine and the test stand to bed. "Pump closed."

"We friggin earned that one," said Hollman with a grin.

I shook Mueller's hand and congratulated him. "Thanks a lot, man," he said. "You saw a good one."

The sun had set, and in the monitors, brush fires flared in the blackness all along the path of the rocket exhaust. While the engineers headed out to the test stand to secure it for the night, I went with a Texan named Cory to help put out the fires. The wind had died down, so there was no danger of the fire spreading, and all we needed to control it were rakes, shovels, and our own feet. Cory drove Rockette and me up close to the nearest of the blazes in an all-terrain vehicle, then got out and headed deeper into the field, stomping out fires with his boots as he went. Joe Allen drove up in another ATV with rakes in the back, and he and I got to work smothering fire with them. The warmth of the fire felt good in the chilly air. "Stay away from the smoke," Joe warned me. "It'll get you. It sure will." The burning grass smelled good, though, mingled with the fresh air. It was a good, clean smoke, not like the urban smog I was more used to.

"Come over here a minute," Allen called to me as we stomped out brush fires. "I want to show you something to watch out for." I crossed to him, and he said, "Fire ants. There's a mound over here. You want to watch out for those. You stand on it, and they'll bite you up real fast." He pointed at some disturbed earth I could barely make out in the dark. "Rattlesnakes are all in moving into their dens right now. No need to worry about them."

Standing out here on this fine night cleaning up the trail of fire left by a powerful rocket engine, it was easy to see why people like Joe Allen and Tom Mueller and the others got addicted to the reek of RP–1 and the electrifying thunder of a sustained, perfectly controlled explosion of fuel and oxidizer. I could think of few things more satisfying in an uncertain world than the *whoosh-BOOM!* of an igniting rocket engine pouring out enough power to hurl a payload into outer space. In sharp contrast to so much of life, no hedging, equivocating, or hand-waving was possible. Rocket engineering either worked or it didn't, and the mere talkers got separated from the doers in that split second between spark and fire.

It seemed clear that SpaceX was well under way toward realizing Musk's goal of becoming the world's first budget space carrier. A half dozen Merlin engines and Falcon 1 rockets were already under construction in El Segundo. Soon, if all went according to plan, the engines and rockets would travel in a steady stream by truck to McGregor for flight qualification testing. From there they would be mated together and shipped to their launch sites. In mid–2005, once the final checkouts were completed—akin, said Musk, to software beta testing—the first Falcon 1 would launch a U.S. Naval Research Laboratory satellite called TacSat–1 from Vandenberg Air Force Base in California. And with that, SpaceX would turn the launch industry upside down.

That was the plan, anyway. Musk spent $7 million refurbishing a pad at Vandenberg's Launch Complex 3-West, and in May, SpaceX engineers raised the first Falcon 1 for a test of launch procedures. The test called for running the Merlin engine for the five seconds that the rocket would normally stay restrained to the pad for final checks before roaring off into space on an actual launch. After a couple of failed attempts in as many weeks, both foiled by the same kinds of problems I saw in Texas—one by a broken sensor, and another by an accidentally-left-closed valve on the pad—the engineers got their dress rehearsal and five-second burn. And then, all dressed up with no date to launch, they waited under Air Force orders for a Titan 4 rocket with a secret military payload on Launch Complex 4-East to take off first so that Falcon 1 wouldn't pose any danger to it as it flew overhead.

After the Titan 4's launch was pushed out to late 2005, Musk got tired of waiting. He decided to concentrate on what was to have been the location of SpaceX second launch. Leaving Tak-Sat–1 back in California, the SpaceX crew shipped the first stage of the second Falcon 1 to the Kwajalein Atoll in the Marshall Islands, about two thousand miles southwest of Hawaii. The second stage and its payload, the FalconSat–2 research satellite built by the Air Force Academy to study plasma in the upper atmosphere, arrived by cargo plane. The crew set up their mission control on Kwajalein Island. A new launch pad went on Omelek Island a few miles away. Both islands were part of the Ronald Reagan Ballistic Missile Defense Test Site, leased by Musk from the U.S. Army.

The islands, Tom Mueller told me, were "really pretty. It's just a tropical paradise. Until you've been there a couple of weeks, and then it's tropical hell." The place was hot and humid, and it offered no chance for visits to or from the families left behind. One engineer quit after a several-month rotation helping to ready the launch site and control center; he'd just gotten married, and either his new marriage or his job had to give.

Finally, in March 2006, after months of work and several false starts, the crew was ready for Falcon 1's maiden space voyage. Space fans around the world watched via webcam as the rocket blasted off, Omelek Island dropped away in an onboard camera view, and then the screen went dark. The engine cut off just thirty-three seconds after ignition, and the rocket fell back to Earth, hitting the coral reef from which the islands were formed only two hundred fifty feet away from the launch pad. The little student-built satellite in its nose popped off, sailed through the air, and crashed through the machine shop next to the pad. Not an auspicious first flight for Musk's fledgling rocket.

Mueller and his crew traced the problem not to mistakes made on the pad or a design flaw, but to an aluminum "B" nut securing a fuel line. The nut had become corroded by the salty spray and fog from the ocean only a few hundred feet from the pad. No longer able to hold a tight seal, it allowed fuel to leak over the top of the engine while the rocket awaited liftoff. After launch, the heat of the engine ignited the spilled fuel, which burned through a pneumatic line charged with helium for activating the engine's valves. Robbed of power, the valves snapped shut, as they were designed to do, shutting off the flow of fuel and oxidizer to the engine, which simply quit.

Understandably, Mueller seemed disappointed as he showed me the wreckage that had been shipped back to El Segundo for analysis by DARPA, which had paid for the launch as part of its own FALCON program. The sad little FalconSat–2, a battered cube, sat on a nearby table instead of hurling through space two hundred miles over our heads as it should have been at that moment. Still, Mueller was confident in his team's ability to bounce back and succeed on the next launch attempt.

Musk himself figured that his customers would give him at least three chances to get it right before they'd stop placing launch orders and he'd be forced to pull the plug on SpaceX. Indeed, they stuck by him through the first crisis; anyone at all familiar with the business of rocketry understood that it was a risky business, especially for a brand-new vehicle. Witness the failures of the first two launches of Musk's nearest competitor, Orbital Sciences' Pegasus, for example.

Even as he and SpaceX were improving the next Falcon 1 rocket to prevent another mishap, Musk was already looking ahead to the Falcon 9. With nine times the thunder and fire of Falcon 1, a Falcon 9 at launch would be an awesome sight to behold. And with its successful debut, Musk's vision for cheap access to space would come to full fruition—Falcon 9 would be able to carry at least five astronauts into low Earth orbit, and perhaps even as many as seven.

SpaceX's *Dragon* space capsule mock-up at the company's shop in El Segundo, California in 2006. *(Photo: Michael Belfiore)*

After getting Mueller's postmortem on Falcon 1 in May 2006, I revisited SpaceX's shop floor and found that Falcon 9's tanks and outer shell were already under construction. Off to one side of the shop sat a full-scale mock-up of the crew capsule, dubbed the *Dragon* (as in "Puff the Magic"), that would ride atop the rocket. The capsule was designed to service the International Space Station under SpaceX's COTS contract with NASA. But Musk also planned to use it to visit another space station, perhaps even a fleet of them, then under construction by a Las Vegas real estate developer with space dreams of his own.

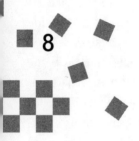

8

BUDGET SUITES OF OUTER SPACE

Robert Bigelow's Forty-Five-Year Plan

Las Vegas looks like an extraterrestrial landscape from the air—miles and miles of empty desert, surrounded by craggy moon peaks. Then the great pyramid of the Luxor Hotel pops into view, and right after that, boom! a line of massive hotels—the biggest in the world, in fact, with thousands of rooms each. The airport sits right up against the Strip like a spaceport on the moon snug against a lunar colony to minimize unnecessary travel over a deadly wilderness.

After landing I pick up my rental car and then drive along the Strip. Each of the monumental hotels has its own water show—particularly outrageous in a desert city—like the Bellagio's eight-acre lake erupting 240-foot water jets in time to show tunes, and the 119,000-gallon waterfall in front of the Mirage. But I'm more interested in a less conspicuous establishment.

On a side road off the Strip, it sits among employment agencies, uniform shops, and other businesses catering to hotel workers: Budget Suites of America. The chain dispenses with the eye-popping displays of its brethren on the Strip, settling instead for a perfunctory neon and flashbulb sign, standard for even the lowliest of Vegas hotels. Affordable weekly and monthly rates, along with laundry rooms and kitchenettes, are the main draws. Some of the windows that look out of the rooms on the ground floor display toys and other knickknacks. People don't just stay here; they live here.

The lobby hosts a few slot machines—again, expected of a Vegas hotel—but no one's playing them when I walk in to check the place out. I overhear a thirty-something man in a security guard uniform apologizing to the grandmotherly type behind the front desk; he's going to be late with his rent that week. No problem, the woman tells him with a kindly smile; just pay up within ten days along with a $10 late fee—or someone will have to come around to change his locks. "I understand," says the man. "Thank you!" An elderly resident comes in to get her mail as they talk. She's lived here for fourteen years, she tells me.

Occupying a valuable niche left vacant by both traditional hotels and apartment buildings, the Budget Suites have made a lot of money for their owner, Robert Bigelow. And though you'd never guess it from these humble abodes, Bigelow has his sights set much higher than the sideshow distractions of Las Vegas. His terrestrial real estate business ticking along nicely without needing a lot of his attention, he now spends most of his time and a considerable chunk of his sizeable personal fortune building Earth's first orbital lodgings. In keeping with his philosophy that quality lodging need not be unduly expensive, he plans to offer rooms for a cut-rate $18.95 million per month. And that really *is* a bargain when you consider that the next cheapest option is $20 million for just a week on the International Space Station via Russian Soyuz. No cheesy water show for this highest of the high rollers; Bigelow's space station will have the oceans of the world on display for its guests as they whiz around the planet at 17,500 miles per hour. Besides just enjoying the view, the station's guests will have plenty of room (three hundred thirty cubic meters, or about as much as a three-bedroom house) in which to learn weightless acrobatics. And every forty-five minutes, when the station's orbit brings it into Earth's shadow, they'll watch the black sky erupt with stars and contemplate their place in the universe.

Bigelow is betting $500 million that he can make his space dream come true. That's about the cost of a typical Las Vegas mega-hotel on the Strip, and more than any other private space venture has at its disposal. Bigelow has also hired a cadre of top NASA and commercial aerospace engineers, and by the fall of 2004, he had already built subscale space station models that were nearly ready to test in Earth orbit. If any entrepreneur could succeed in launching the planet's first commercial space station, it was Bigelow.

Getting guests to the front door—or airlock—was Bigelow's greatest challenge. He could get his space station components into orbit using existing unmanned commercial rockets, but no commercial orbital spaceships yet existed. That's why, even as Brian Binnie was landing *SpaceShipOne* after its triumphant X PRIZE–winning flight, Bigelow was launching his America's Space Prize—$50 million for the first non-government venture to send people to orbit. After an unsuccessful attempt to get NASA to front half of the prize money, Bigelow committed to paying out the full $50 million himself rather than waste any time trying to find other sponsors; his strict schedule called for his space station to be fully operational by 2010.

From the North Las Vegas street it occupies, Bigelow Aerospace, or BA, as it is known to those who work there, looks like just another nondescript industrial development in the area—the beer distributor across the street, for instance. That is, until you drive past the staid BA logo out front and into a parking lot bordered with razor wire–wrapped chainlink fencing. The lot sits atop a hill overlooking the Las Vegas Valley and the distant Strip. There's nary an office building or warehouse in sight—just a small prefab building across the parking lot from the gate.

On my first visit, a couple of beefy guards in desert fatigues, 45s on their hips, check my ID, take my backpack, and lock it up for me. They seem like nice guys, but they don't have to behave otherwise to make it absolutely clear that I am to do exactly as I am told. Give up any recording devices I might have, along with a cell phone that might be capable of taking pictures, for example. I dutifully relinquish my tape recorder. On that first visit, I am too nervous to notice something that blares out at me the next time: each security officer wears a black shoulder patch bearing a silver and gold alien face.

After I get cleared through security, Bigelow meets me down the hill from the guard post. He is a fit sixty years old, with a full head of salt-and-pepper hair and a neatly trimmed mustache. He wears immaculately pressed dress trousers with a colorfully patterned button-down shirt and spotless white sneakers. With one of the guards shadowing us at a discreet distance, he shows me around the grounds of his fifty-acre R&D facility. His manner, though subdued and unpretentious, carries an air of utter self-confidence. "It's a gamble," he admits to me of his project, which, besides being the most expensive private space venture ever undertaken, is certainly the most ambitious. "It's a huge gamble." But the idea of his big gamble clearly pleases him—as befits someone whose hometown is Las Vegas. Of course, he also had deeper reasons for undertaking this particular gamble.

While Robert Bigelow was growing up in Las Vegas in the 1950s, nuclear test explosions at the Nevada Test Site sixty-five miles away lit up his neighborhood at night with a pale, artificial sunlight. He and his classmates often turned out to watch day-

Robert Bigelow. (*Courtesy of Bigelow Aerospace*)

time tests from their schoolyard, and many of their homes had hairline cracks in the walls and ceilings—the result, Bigelow believes, of nuclear earthquakes. Some people, including members of Bigelow's own family, spoke of an even more powerful force lighting up the skies over Las Vegas. Bigelow was nine years old when his grandparents told him a story that was to shake him at least as much as those nuclear blasts.

Late one night in 1947, Bigelow's grandparents were driving down a desert highway when they saw a glowing red object in the distant sky moving toward them impossibly fast. They pulled over to watch, and in seconds the thing filled the windshield. They ducked down, certain they were about to die. But an instant before impact, the object veered sharply and sped away. They believed it was a vehicle of some kind, but it wasn't like any vehicle they'd ever seen. No craft in existence, no airplane, no rocket could turn on a dime at high speed like that. "When people in your family have an experience that's earthshaking to them," Bigelow told me later, "then you begin to take them seriously." Bigelow heard more stories of UFO encounters like this one, and the more he heard, the more certain he became that humankind was being visited by beings from another world.

That got him thinking about humankind's place in the universe. If we were being visited by aliens, then clearly they were far more advanced than we. We hadn't even begun to leave the atmosphere, and these beings were traversing interstellar distances. What's more, they were interacting with us without the majority of us even knowing about it, which also implied an advanced technology. They must be benign, Bigelow thought, otherwise they would have had their way with us long ago. He figured they must be studying us, like anthropologists trying to observe a primitive tribe without disrupting their way of life. Picture a valley, he told me. We're completely enclosed by mountains, and we can't get out. All we know is what we can see. Our campfires. Our simple weapons and tools. We have no concept of the greater universe around us. We also believe we're the most advanced, most powerful beings around.

The fact that they've kept themselves relatively well hidden, insulating us from the shock of seeing ourselves reduced to the status of primitives, is another reason Bigelow believes the aliens are kindhearted, that they want us to be free to pursue our own destiny: either blow ourselves to bits or join the great galactic community. As Bigelow saw growing up, we stood at the crossroads; at the same time we were taking our first tentative steps into space, we were also developing weapons so powerful that they could wipe us off the face of the planet. Bigelow thought deeply about these issues as he grew older, and he chose optimism over pessimism. The best of human nature could win over the worst, he believed. Our destiny lay in space. Only there would we begin to develop the technologies we needed to take our rightful place among the other advanced species of the galaxy.

Gravity, or its absence, was the key. "Working in space you can spawn architectures of things that are not possible under gravity influence—just like you couldn't if you had a periodic table on some other planet that had only half of the elements that Earth has." Under the influence of gravity, Bigelow told me, "You're not dealing with a level playing field. It's not fair. You're starting off on something that has got two feet back from everybody else who has far more elements to obtain."

By the time Bigelow was fifteen years old, he had decided on his life's mission: to help humankind establish a beachhead in space, to help get us that level playing field on which to meet our cosmic destiny. And although he didn't know then exactly what form his vision would eventually take, he did know he would need money, and a lot of it. How to go about getting that money was a no-brainer. His father was a real estate broker, and his grandfather (he of the UFO encounter) owned half a dozen small apartments that he rented out. Bigelow grew up next door to his grandfather, so real estate became every bit as much a part of his heritage as speculations about other worlds. He studied real estate and banking at Arizona State University. After he graduated in 1967, he became a real estate broker and then began buying his first small rental properties.

His big break came in 1970, when he got a bank loan to finance his first construction project, a forty-unit apartment building. He was then just twenty-five years old, and at first no one would lend him the $320,000 he needed. But patience and dogged persistence paid off. The twenty-sixth lender he approached gave him the money, and he was off and running on what was to become an extremely successful construction career. During the 1970s and 1980s, he built dozens of apartment buildings, offices, warehouses, and a few motels in the Las Vegas area. In 1988, he saw a niche to be filled in the area of extended-stay units, and he founded Budget Suites of America. The Suites are more like apartments than hotels, as far as Bigelow is concerned: "It's just that I was willing to accept rent in increments less than a month. So if they wanted to pay by the week, that's fine. It just cost them a little more for that service."

At about the time he founded Budget Suites, Bigelow decided that he finally had enough money to begin seriously investigating the UFO phenomenon that had gotten him so interested in space travel. He began financing research into UFOs and other unexplained phenomena to the tune of several million dollars, and in 1995 he founded an organization called the National Institute for Discovery Science (NIDS). In the nine years of its operation, until Bigelow put it on hiatus in 2004, the organization employed scientists and other researchers for investigations into "aerial phenomena" (i.e., UFOs), unexplained animal mutilations, and "other anomalous phenomena," according to its Web site. For a time, NIDS was the Federal Aviation Administration's only officially recognized point of contact for the reporting of UFO sightings.

Although he never learned anything conclusive about who or what was causing the numerous UFO sightings he investigated, Bigelow remained convinced that they were the result of visitations from extraterrestrials. And even as he amassed a greater and greater fortune from an ever-growing real estate empire, he kept his focus on his childhood dream—to help humankind expand beyond planet Earth. But he kept his real motivations a secret, even from his wife. "She never knew," he told me later. "Because it's possible that that kind of dream would never happen."

Finally, when Bigelow was fifty-five years old and worth hundreds of millions of dollars, he picked up the April/May 1999 issue of *Air & Space Smithsonian*. There, in an article by Marcia Dunn, he read about a radical new space station under development by NASA. The new design not only improved on traditional designs but was cheaper to boot. It seemed like a great idea to Bigelow, and when Congress killed the program the following year, Bigelow knew that his opportunity had arrived.

. . .

Bigelow led me to a door in the otherwise featureless wall of one of two multistory buildings on the Bigelow Aerospace grounds. Inside it was cool and dark, in contrast to the midday desert summer outside. An expanse of unmarred white floor led to the back of the building a few hundred feet distant, where two watermelon-shaped structures loomed out of the darkness. The one facing us was draped with an American flag. The structures glowed flawlessly white, like the floor and Bigelow's sneakers. These were full-sized mock-ups of Bigelow's dream baby: the BA–330 space station module, one of which would constitute Bigelow's first space station.

Bigelow planned to put self-contained BA–330s into production like automobiles, albeit very expensive ones. He figured he could build them for around $100 million each, gang them together into orbiting "space complexes," and then lease them out for a month to a year at a time to anyone who wanted to establish a presence in space—mainly national governments and corporations. At forty-five feet long and twenty-two feet in diameter, the BA–330 would have more habitable volume than anything ever launched into space in one piece. These two mock-ups, linked end to end by a docking module as they might be in orbit, enclosed an interior volume larger than that of the then-current configuration of the International Space Station.

I found myself tearing up. "That is awe-inspiring," I breathed. This guy was actually building real hardware, not just turning out the computer-generated concept images that aerospace companies seemed to love so much. Bigelow made no comment to my little exclamation, though this was surely the reaction he'd hoped for. He had me sit with him on the metal chairs just within the entrance and put on white paper booties so we wouldn't scuff the white floor. Then, with the security guard standing at attention and watching us from the doorway, we headed in.

We were in Building B—eighty thousand square feet of open work and display space. At the time of my visit, Bigelow was gearing up for tours of the building by local schoolchildren, but very few outsiders had yet been inside. We walked past a seven-foot-tall bronze sculpture of a bald eagle, with wings outstretched, landing on a rocky outcrop. Its purpose, Bigelow told me, was to remind visitors that his was, first and foremost, an American enterprise. Walking through the beginnings of an exhibit for the school tours, we reached the foot of the modules and climbed a steep metal stairway to go inside.

The relatively spacious interior of the BA–330 comes from a most unusual property for a space station—it is inflatable.

The idea of an inflatable space station may conjure an image of an easily punctured party balloon–like structure, but the TransHab design is actually *more* durable than the rigid aluminum structures that ultimately made it to the International Space Station (and every other government space station ever built). Tests of the BA–330's MicroMeteroid and Orbital Debris (MMOD) shield, the station's outer shell made up of five layers of carbon-fiber composites, showed that it could shrug off five-eighths-inch projectiles fired at 6.4 kilometers a second. That's six times faster than an M16 rifle bullet, which should reassure future astronauts. International Space Station astronauts enjoy no such protection from micrometeoroid strikes.

The space station module's three main layers, including the MMOD, remain flexible until pressurized, allowing the space station module to ride to orbit with its hull wrapped around its aluminum-frame core section. This gives the design another important advantage: since it launches as a tightly wrapped package and inflates to full size only after reaching orbit, a much larger space station can be launched with the same size rocket booster that would lift a smaller rigid structure. Current boosters simply aren't big enough to launch a rigid structure as roomy as a fully expanded BA–330.

Straps will hold the BA–330's soft and flexible hull in place for launch, but after it arrives in orbit, explosive bolts will blow the straps off, and then the station will inflate with air from tanks in the core. Only fifteen feet around at launch, the thing will expand into a bulging twenty-two feet in diameter. Solar panels will unfold to catch the sun from noninflatable air-lock-and-docking adaptor units at both ends. At that point the walls of the station will be as hard as concrete and much stronger than aluminum.

After controllers on the ground verify that all systems on the space station are go during a checkout period that could last several months, the station's crew will arrive in a smaller spaceship to dock and enter through the airlock. Weightless, the astronauts will float through the airlock into a comfortable but largely empty environment, much like the one I found myself in after climbing the stairs on Bigelow's mock-up. The four longerons, analogous to poles in a tent, that define the core of the BA–330 run through the center of the station, from one airlock to the other. Along with the life support, communications, and other systems, the core contains food, water, and all the gear the astronauts will use in flight, along with removable panels like the walls of terrestrial office cubicles. The crew can use these panels to delineate as many as three levels of space in which to live and work in the station.

Although the metal mock-ups I toured at Bigelow Aerospace had no windows (the real station will showcase spectacular views of planet Earth and the stars, naturally), they made me feel not at all claustrophobic. I entered onto a loosely woven netting of nylon straps that let me peer down into a lower level. A comfortably high ceiling of the same material let me see into an equally spacious upper level. On the actual station, the crew will use the removable partitions from the core to create a galley area, as many as six small bedrooms, and perhaps a play area where guests will have room to enjoy weightlessness properly (or get properly spacesick).

A so-called multidirectional propulsion bus (MDPB), launched separately from the crew, will turn the space station into a true spaceship, giving it the ability to maneuver to another orbital inclination to dock with a sister station, meet an arriving passenger ship, or even to leave orbit, say for a cruise to lunar orbit.

All of this basic architecture was worked out by senior NASA engineer Dr. William Schneider at NASA's Johnson Space Center in Houston. After Schneider and his team of engineers got started on the project in 1997, they quickly gained political allies at NASA, including then-administrator Daniel Goldin, who hailed it as "one of the major breakthroughs in the space program," the kind of breakthrough, in fact, that only "comes around every ten or twenty years." After doing time as part of a planned Mars mission that never left the ground, Schneider's Transit Habitation, or TransHab, moved over to the International Space Station, where it was being groomed for use as the International Space Station Habitation Module. Donna Fender, TransHab's project manager, figured that NASA could complete an inflatable Habitation Module for Space Station for about $200 million. That's $100 million less than Boeing charged NASA for its Unity Node 1, launched to Space Station in 1998. Space Station program manager Randy Brinkley thought that made TransHab sound like a pretty good idea. "Clearly we would prefer to go with TransHab, all things being equal," he said in 1999.

But Boeing made sure that all things *weren't* equal. Every time the TransHab team released a new estimate for what the inflatable Habitation Module would cost to build in house, Boeing fired right back with a correspondingly low bid for its competing aluminum structure. To a NASA official quoted anonymously in the April/May 1999 issue of *Air & Space Smithsonian*, that sounded like crap. The first estimates from Boeing weren't "anything like" TransHab's relatively low cost, the official said, and "I'm not sure how real those figures are." Nevertheless, those figures formed the basis for the decision on whether to use TransHab or Boeing's traditional design for the Habitation Module.

Congress made that decision with House of Representatives bill H.R. 1654, which set NASA's budget for fiscal years 2000 through 2002. The bill specifically prohibited NASA from using any money to build TransHab because, in the words of Congressman James Sensenbrenner, chairman of the House Committee on Science, "TransHab costs more, and we simply don't have the money." Sensenbrenner's report accompanying the text of H.R. 1654 listed the cost for Boeing's aluminum Habitation Model at $186.9 million, comfortably lower than NASA's estimate for TransHab, which by then had grown to $250 million. After H.R. 1654 became law, NASA disbanded the TransHab team. Some of the engineers remained with NASA and moved to other programs; others retired. Meanwhile, Boeing's Habitation Module never got built either, and the space station itself remained chronically short of living space, able to accommodate only three astronauts at a time instead of the six or seven originally planned for.

Bigelow thought he could complete TransHab simply by applying standard business practices to the problem of building a space structure, just as he would for any terrestrial construction project. "I've put together many, many projects involving a lot of money and a lot of people," he said. "I'm used to doing things pretty darn well on budget and pretty darn well on time." He cut a deal with NASA to acquire the exclusive right to develop the TransHab design, and he planned to build his modules with as many low-cost, off-the-shelf components as possible. He would steer clear of the exorbitant satellite launchers of the major launch providers in the United States and instead send his modules to orbit on cheaper Falcon rockets from SpaceX and converted ICBM Dneper rockets from Russia.

To gain access to former TransHab engineers still at NASA, Bigelow entered into a Space Act Agreement (a NASA program for sharing expertise with private enterprise), which allowed him to borrow them from their usual duties for consulting visits to Bigelow Aerospace. TransHab inventor William Schneider had by that time retired from NASA to take a teaching gig at Texas A&M University. Bigelow tracked him down and invited him to Las Vegas to see what he had built so far. "And God," Schneider said later, "when I walked in here, boom! It was mind-boggling, because this is the vision that I really wanted. Here's these things, all sitting there, and of course some of them are mock-ups, but the rest were inflatable, and I said, 'Man, he's serious. He's not playing around.'" Schneider cut his teeth at NASA during the Apollo years. Some of his designs are still on the moon today, like the cosmic rickshaw used by the *Apollo 14* crew to tote equipment across the lunar surface. In his seventies when I met him, he still exuded a youthful exuberance, and in contrast to the younger engineers' more casual attire, he wore a sport jacket and tie to work. He was absolutely delighted to be able to pick up his project where NASA had left it. "It's kind of like you want to see your child grow up to maturity," he told me, "not be stopped in its adolescence." With visits to Bigelow Aerospace every other week or so with his former TransHab engineers to consult with Bigelow's full-time engineers, he was well on his way toward realizing that desire.

Machinists and technicians labor in the forty-thousand-square-foot Building A, assembling small-scale test versions of space station modules. Like Building B, where the mock-ups live, the space is vast and windowless. Big enough, in fact, to house a prefab two-story office building tucked away in one corner like a storage cabinet in a garage. Bigelow's engineers work there on the ground floor. As they complete their designs they transmit them to the computer numerical control (CNC) milling machines on the shop floor, and they watch from their desks as the technicians turn computer code into shining steel and aluminum parts.

Engineer Edwin Lardizabal, a talkative Filipino, heads the structural design department. He worked on passenger jet airframes at Boeing for ten years before he lost his job amid the post–9/11 downturn in commercial aviation. Life support chief Richard Chu, a Chinese American who helped design the life support system on the International Space Station during his twenty-five years at NASA, exults in his role at Bigelow Aerospace. At NASA he was but one of an army of environmental control and life support systems (ECLSS) engineers. Here, he *is* the ECLSS department. Based on the International Space Station design, Chu's new life support system will be a generation better.

Bigelow spends a lot of time out of his office in the prefab building and out on the shop floor, where he can stay intimately involved in every aspect of the day-to-day work at the plant. He commands respect from his employees; in an era when it's common for workers and managers alike to address each other by their first names, Bigelow is known around the plant as Mr. Bigelow, or simply Mr. B. Yet, he's also well liked. "If you're an employer and you're disliked," he tells me, "sooner or later you're going to be ineffective."

Bigelow has little patience for intangibles. In the summer of 2004, he and his engineers met to decide whether to use NASA's JPL for vibration tests of test modules. As Bigelow's patent attorney, Franklin E. Gibbs of the law firm Wang Hartmann & Gibbs, later recalled, Bigelow cut short the discussion, took the stunned engineering team out to the airport, put them on his private jet, and flew them out to Pasadena to base their decision on firsthand observation rather than abstract ideas. They ended up using the facility.

On another occasion, some of the engineers were deadlocked about what size to make the hardware that held the straps of a test module's restraint layer in place. The restraint layer, just beneath the outer skin of the module, is what keeps the innermost layer, a plastic air bladder, from bursting when it is under pressure. It's made up of numerous overlapping straps of Vectran, a synthetic fiber twice as strong as Kevlar, five times as strong as steel, and ten times as strong as aluminum. Just as important to the integrity of the space station as the strength of the straps themselves is the means by which they stay attached: with metal rollers and so-called clevis fittings at each end of the module.

The engineers had been using one-eighth-inch rollers, but some of them wanted to make the rollers bigger—three-sixteenths of an inch—to make doubly sure they stayed in place. Gibbs later recalled the scene: "We've got a room full of engineers, and everybody is worried about figuring it to the nth degree, and Robert just says, 'Wait. Build it. Let's see what it does.'" Bigelow called the manufacturing manager into the meeting. "Build both of them," Bigelow told him. "I want a dozen of these ready after lunch." The engineers came back to find a dozen new rollers of each size laid out on the table ready for their inspection. The result: they went with the larger (and more secure) three-sixteenth-inch rollers.

Putting preliminary designs into physical form isn't exactly the norm in aerospace engineering, and it took some getting used to for Schneider. At NASA, Schneider explained to me, "We would sit down and do all the engineering first before we ever cut any metal, period." When Schneider joined Bigelow Aerospace as a consultant, he was amazed by Bigelow's insistence on producing a machined part for every revision of a design—and the skill with which Bigelow's machinists turned them out in short order. Schneider's first task at Bigelow Aerospace was to fly in twice a month for a couple of days at a time to help Bigelow's staff engineers with the design of the space station's windows. Schneider told Bigelow he planned to "go ahead and design it right the first time" before spending a lot of time machining parts that weren't ultimately going to be used. But Bigelow would have none of it. "Well, I'm not used to that," he said. "I have to keep these guys busy out there because I don't want them sitting around. So design it the best you can in that two-day period, and when you come in the next time, we'll have one sitting on your desk."

"And it absolutely blew me away," Schneider told me, "because he did." Schneider would set to work with a felt-tipped pen revising the part on the spot, drawing on it to indicate what sections he wanted to cut away, and Bigelow would always stop the machinist from putting it back in the Computer Numerical Control (CNC) machine. "No, no," Bigelow told him. "We'll make a new one."

"So we redesigned it," Schneider told me, "and the next time I'd come, *Shoomp!* it had evolved to the thing I and some of the other fellows here designed. In about four of those periods, we had the window concept that could take the load across it very gently and be relatively lightweight."

Schneider feels that his brainchild is in good hands with Robert Bigelow. He likens Bigelow to another hugely successful aerospace entrepreneur who cleaned up in Las Vegas real estate: Howard Hughes. Hughes was indeed an inspiration for Bigelow. As Bigelow remembers it, Hughes almost single-handedly pulled Las Vegas out of a recession in the late 1960s, right around the time Bigelow graduated from college. "He was probably the first influential high-profile, legitimate corporate mogul who was taking Las Vegas seriously," Bigelow told me. But Bigelow himself resists the comparison. "I'm no Howard Hughes," he says. "He was an extraordinary man. He was a genius. He was also kind of a peculiar fellow."

Bigelow's approach to his space station project is anything but peculiar, spaceship life support expert Taber MacCallum told me. MacCallum heads Paragon Space Development, life support system contractor for both SpaceX and Rocketplane Kistler. A visit to Bigelow Aerospace and conversations with Bigelow's engineers conquered his initial skepticism. "They're taking a very down-to-earth approach to what they're doing in terms of building and testing," he told me of Bigelow Aerospace. MacCullum told me he felt Bigelow's approach was aggressive, yet very concerned with safety, like Burt Rutan's work on *SpaceShipOne*. Space industry analyst John M. Logsdon, director of the Space Policy Institute at George Washington University, agreed that "the basic technology" behind Bigelow's space stations "is likely to work." Bigelow's big problem, said Logsdon, was "whether there's a transportation system that can get people or things, or both, up there." Logsdon was more familiar with problems in space transportation than most; he served on the *Columbia* Accident Investigation Board, the official body that sought the cause of space shuttle *Columbia*'s fatal disintegration on reentry in February 2003.

Before the *Columbia* disaster, Bigelow planned to supply his space stations with food, air, and water and to send crews up with Russian-built-and-launched Soyuz space capsules. "They were ready, willing, and able to deliver those Proton rockets and those Soyuz vehicles," Bigelow told me of the Russian aerospace firm RSC Energia. After *Columbia* went down, though, Bigelow suddenly found himself competing directly with NASA for Soyuz flights; with the shuttle out of commission for an indefinite time, the Soyuz was the only way for NASA's astronauts to visit the International Space Station.

But the X PRIZE's success in fostering private suborbital spaceflight opened a possible way out of Bigelow's quandary. Why not offer a prize for the first private *orbital* spaceflight? He had the money to front such a prize himself, without the search for sponsors that had consumed so much of Diamandis' time and that had ultimately forced the X PRIZE to have a deadline. Bigelow liked the idea of a deadline, though; he could set it to coincide with the planned launch of his first space station. Bigelow sketched out the details of what he was to call America's Space Prize for me over the phone on October 4, 2004, just hours after Brian Binnie landed *SpaceShipOne* following his X PRIZE–winning flight. He released the full list of ten rules for winning the $50 million prize when I met with him in Las Vegas the next month:

1. The Spacecraft must reach a minimum altitude of 400 km (approximately 250 miles);
2. The Spacecraft must reach a minimum velocity sufficient to complete two (2) full orbits at altitude before returning safely to Earth;

3. The spacecraft must carry no less than a crew of five (5) people;

4. The Spacecraft must dock or demonstrate its ability to dock with a Bigelow Aerospace inflatable space habitat, and be capable of remaining on station at least six (6) months;

5. The Spacecraft must perform two (2) consecutive, safe, and successful orbital missions within a period of sixty (60) calendar days, subject to Government regulations;

6. No more than twenty percent (20%) of the Spacecraft may be composed of expendable hardware;

7. The Contestant must be domiciled in the United States of America;

8. The Contestant must have its principal place of business in the United States of America;

9. The Competitor must not accept or utilize Government development funding related to this contest of any kind, nor shall there be any Government ownership of the Competitor. Using Government test and launch facilities shall be permitted; and

10. The spacecraft must complete its two (2) missions safely and successfully, with all five (5) crew members aboard for the second qualifying flight, before the Competition's deadline of January 10, 2010.

Like both of the X PRIZE flights, the first America's Space Prize flight may carry just a pilot and ballast to equal the weight of the passengers. But Bigelow insisted that the second America's Space Prize flight carry five actual people, not just ballast. The wording "subject to government regulations" was significant, though; the FAA had so far licensed only people necessary to the operation of the craft—i.e., pilots—to fly on commercial spaceflights. That's why, even though Burt Rutan hinted that he might ride in the back seat of *SpaceShipOne* on Binnie's X-Prize–winning flight, ultimately the ship carried only Binnie and the weight equivalent of two passengers. It would be up to Virgin Galactic to win approval from the FAA to fly actual paying passengers. Understandably, Bigelow would like the ship that won his prize to come preapproved for passenger use.

Bigelow fills me in on the business plan for the ships that will service his space stations as we sit in the second-floor conference room in Building A. "There's more of a financial benefit than this $50 million by far," Bigelow explains. Not only the winner of the prize but also runners-up will get to compete for BA contracts.

Bigelow gets up from his seat, looking at the wall over my shoulder. "That piece of tape up there has been sticking there since I've been talking to you." He walks around the table to peel off a bit of transparent tape from the otherwise bare wall behind me. Then he crosses to a whiteboard at the end of the conference table. When he turns his back to sketch out diagrams to illustrate his plans, I notice a single long hair stuck to the back of his jacket. It stands out, I realize, because his clothing—indeed, every aspect of his appearance—is without flaw. *Everything* Bigelow does, it seems, is studied, meticulous, carefully considered.

Whoever can service the space station, Bigelow explains, will get $32 million per flight from Bigelow Aerospace. That's in 2004 dollars; the rate will go up 2.5 percent each year from now on to adjust for inflation. Bigelow will offer a contract for six flights over a 12- to 18-month period, starting in January 2010. That contract will be worth $192 million, again in 2004 dollars, which works out to $200 million in 2010 dollars. "In addition to that," Bigelow tells me, "we want to option for twenty-four more flights over the course of four years, which is from 2010 through 2013. Now, that's about $800 million by the time you add the 2 percent inflation. So what this is, is a billion-dollar package. What do we do with this?" He writes a figure on the whiteboard. "That's a ticket price per person. This buys somebody one week on station." The number gets my attention: $7.9 million, or $12.1

million less than the Russians charge for the same amount of time on the International Space Station.

(In early 2007, Bigelow revised that figure to $18.95 million per month. That was in 2012 dollars, Bigelow's new launch target date. Most of that fee would go toward transportation to orbit, and it was even less per week than Bigelow's original price. To buy an entire additional month on station, customers would have to pay only $2.95 million extra.)

Although the space stations will welcome tourists, Bigelow sees his main market as foreign governments tired of waiting to fly their astronauts to the International Space Station on other governments' ships, and major corporations wanting to use the unique properties of microgravity for product research and development.

That night, *Popular Science* staff photographer John B. Carnett arrives with an SUV full of equipment and two assistants. Bigelow has never before allowed himself to be photographed by the media; he wants to be able to go out in public without being recognized, and he's always worked hard to keep his personal life separate from his professional life. But he's agreed to have his picture taken for the article I'm writing. It's a big moment for him, and he's a little nervous. He tries to keep things light by joking that maybe Carnett would like to shoot him looking up at the sky and pointing at UFOs.

Carnett spends a couple of hours composing the shot he wants and setting up his lighting. He wants Bigelow in a folding chair, writing on a notepad, in front of a pair of test modules undergoing leak tests. But Bigelow thinks he'll look silly posing. He wants to know why he can't be photographed candidly during the normal course of his work. Carnett tries to explain that he won't get the best portrait that way, but to no avail. Bigelow walks out of the frame without giving him the shot and heads back to his office with Schneider, who's come along to watch. I join them as they cross a deserted section of Building B. Bigelow tells me how he'd like to be photographed—just as he is now, walking with Schneider, "pop, pop, pop," without even breaking his stride. In other words, Carnett should work around him, not the other way around. I can see it: the engineer and the space mogul discussing the invention that could change the world (or at least life in space), moving through a black, empty space from one pool of overhead light to another.

Carnett finally gets his shot as I sit with Bigelow in his small uncluttered office back in Building A. That's when Bigelow tells me he thinks it's more than likely he'll lose his $500-million gamble. "But you know," he says, "the faint of heart never won a fair maiden, never won wars." Besides, he believes that what he's doing is good for the country, whether he succeeds of fails. The United States is falling behind the rest of the world in space. If this country has any chance at all to catch up, it'll be the private sector that does the job. As Bigelow and I talk, Carnett photographs him through the window in his office door. The shot that goes in the magazine captures the moment perfectly. There's Bigelow's face, partially obscured by the window frame, half in, half out of the public eye. Almost every available space on the wall behind him is covered with engineering diagrams, notes, and descriptions of space station hardware; they're Bigelow's thoughts made manifest.

Ultimately, for Bigelow, it all comes down to inspiration. He was inspired as a kid to get into space, and he hopes to inspire a new generation. "Where's the inspiration in America is really what it boils down to today. If you asked fifty people or five hundred people, 'What is America's inspiration today?' What the hell would they say? To win the war in Iraq? How the hell can that be an inspiration? That doesn't create a dream in some kid's mind. An inspiration has to be something you carry with you 24/7."

In July 2006, Bigelow took a big step toward realizing his dream when a Dneper rocket lofted a one-third-scale test version of the BA–330 called *Genesis I* into orbit from the ISC Kosmotros launch complex in Russia. The fifteen-foot-long ship successfully deployed its solar panels, inflated to its full eight-foot diameter, and began beaming telemetry and video back to mission control in Las Vegas. "It felt like becoming a parent," Bigelow said of the successful test.

Although far too small to accommodate people, *Genesis I* did the next best thing: it carried photographs, small toys, and other mementos contributed by BA employees. Cameras inside the craft sent back videos and still images of the objects whirling around the interior in a kind of psychedelic space show while cameras outside the ship sent back spectacular views of Earth from orbit—some of which BA posted on its new Web site. The new site also invited the public to participate in the launch of *Genesis II*, which would carry anyone's photographs or small objects for $295 each. Your money back if you didn't see your smiling mug or tchotchke sailing through space at www.bigelowaerospace. com within ninety days. Bigelow had come a long way from his days as a semirecluse. "We are out of the closet, that's for sure," he said.

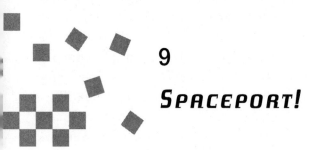

9

SPACEPORT!

Launched by air from the belly of a jet-powered mother ship. Stacked up on the ground in stages and then booming into the air and dwindling out of sight in seconds flat, squirting fire and a smoke trail that drifts slowly on high-altitude winds. Lifted fifteen miles high by a giant balloon to cut loose and blast into space. No matter how they get there, spaceships need a place to launch from. Someplace with plenty of room for takeoffs and landings, lots of open sky with no interference from mere airships. Room enough to crash without killing anyone on the ground or destroying buildings. Patient ground crews who understand the meaning of hurry up and wait—wait for winds to die down, wait for lox tanks to fill, wait for countdown holds to clear. Rocket ships need spaceports, and ordinary airports, with their close proximity to major metropolitan areas, scores of daily takeoffs and landings by airplanes with schedules to keep, and swarms of people coming and going just won't cut it.

Even before the first spacelines send the first paying sub-orbital passengers aloft, localities across the United States and around the world are doing their best to attract the emerging commercial spaceflight businesses. To some places with plenty of open space—in the middle of nowhere, in other words, and typically left behind by traditional businesses—spaceports look like economic salvation. Locales in Singapore, the United Arab Emirates, Sweden, and Scotland all hope to host spaceports in the near future. In the United States, New Mexico, Wisconsin, Florida, California, Virginia, and Texas all play host to embryonic commercial spaceports.

With their existing infrastructure and/or active influential supporters, the spaceport plans of California, New Mexico, and Oklahoma seem to have good prospects for success. But only one of them has hosted manned commercial spaceflights—the air-port-turned-spaceport at Mojave, California. And, ironically, Mo-jave gets perhaps the least support from its state. In part this is due to the state's relative prosperity and cash-rich government aerospace projects stealing focus, as well as its multitude of other thriving industries like entertainment and information technol-ogy. With the sixth largest economy in the world to look after, California officials simply have too much on their plates to de-vote much time or many resources to a speculative industry that might or might not bear fruit. Then, too, regular breakthroughs in aerospace just seem like business as usual to anyone at all familiar with the Mojave Desert—hardly something that needs special nurturing.

The town of Mojave is only a hundred miles north of Los An-geles, but it might as well be on another planet. To get there from the Los Angeles International Airport, you take the 405 freeway through the San Fernando Valley and then get on Highway 14, the Antelope Valley Freeway. As you come over the San Gabriel Mountains into the Mojave Desert, the landscape turns arid and the temperature rises. In summer the moisture seems to evaporate right out of your body as the sun blasts down on the roof of your car. Without thinking much about it, you can drain a bottle of water in a few swallows and then reach for it again because you're still thirsty.

On the other side of the mountains, in the Antelope Valley, is the city of Palmdale, where NASA's space shuttles and the abortive X–33 were built. Twenty miles farther on, past the town of Lancaster, you pass Rosamond and the turnoff for Edwards Air Force Base, where Chuck Yeager broke the sound barrier and the space shuttle sometimes lands after missions. After another twenty-minute drive down a long flat stretch of highway through desert sand punctuated by Joshua trees and strewn with tumbleweeds, you at last reach the town of Mojave.

It doesn't look like much from the highway—railroad tracks, long lines of boxcars and tankers, a string of motels and gas stations, fast-food joints and greasy spoons. It would be easy to drive right through and never give it a second thought but for the sign on the outskirts of town proclaiming it the home of *SpaceShipOne*. The bright graphic of the little spaceship touching down is the only spot of color in the otherwise drab landscape. Power-generating windmills spin in the foothills of the Tehachapi Mountains on the other side of town. The most prominent structures are the parked airliners at the airport, stored there because the desiccating desert air preserves them.

The airport on the east side of town covers more area than the clutch of houses and other small buildings that constitute the rest of Mojave. The Southern Pacific Railroad laid out the streets and lots back in 1876 to service its rail line between Los Angeles and San Francisco through the Tehachapi Pass. Mojave had a hard time taking root there among the Joshua trees and scrub brush; its major buildings burned down three times in the 1880s. But two or three hundred railroad workers and some entrepreneurial types hung on. J.W.S. Perry was one of the latter, building gargantuan wagons with wheels as tall as a man and pulled by twenty-mule teams. The wagons hauled borax, a mineral used in metalworking and enameling and as a cleanser and food preservative, to the railroad depot from mines a fifteen-day journey away. A plaque on Highway 14, Mojave's main drag, is the only sign left of Perry's place, which is now occupied by an auto shop. In the 1890s, gold was discovered in the nearby foothills, though that doesn't seem to have had a lasting effect on the town's fortunes.

At the time of the X PRIZE flights in 2004, Mojave looked like a relic from the Old West. Its largely empty windswept streets, the occasional boarded-up building and the fading 1950s-era White's Motel proudly advertising TV in its rooms all contributed to the impression that the place was struggling to keep ahead of the desert sands that sought to claim it. A highway bypass, an extension of Highway 58, had just been built, routing much of the semi-truck and tourist traffic passing through the area around it. Mojave started life as a place on the way to someplace else, a stopover rather than a destination in itself. It has continued thus. "How many people live in Mojave?" asks a Mojave joke. "About half of them," goes the answer.

The airport exists almost as a world apart from the town, with workers buzzing through and disappearing to work in the World War II–era hangars and bungalows, emerging only for flight tests and lunch at the Voyager Restaurant, the diner located in the single-story administration building just below the old control tower. The U.S. Marines took over a dirt landing strip in 1942 and built up the present-day airport as a training center for aircraft carrier pilots. Shut down after World War II, the airport opened again in the 1950s for the Korean War and closed in 1959. Then, in 1972, an enterprising soul named Dan Sabovich established the place as a civilian airport, the Mojave Airport District—later renamed the East Kern Airport District after the county in which Mojave resides.

After a shaky start, during which the airport's biggest customer was a farmer who dried grapes on the airport's silent runways, business gradually picked up. In 1974, Burt Rutan's Rutan Aircraft Factory moved in, and from there things began to roll, albeit probably not as rapidly as Sabovich would have liked. In 1981, the National Test Pilot School arrived, and it quickly became a premier training ground for pilots from all over the world; of the top seven test pilot schools in the world, it was the only one not under military control. In 1982, Rutan founded Scaled Composites. In 1986, Rutan's brother, Dick, and Dick's girlfriend, Jeana Yeager, and their volunteer crew built *Voyager* at the airport and flew it nonstop around the world from Edwards Air Force Base. After the Voyager Restaurant moved in below the control tower, it quickly became the place to meet at the airport, filled to capacity during the lunch hour, when diners could watch the flight line from the big picture windows at the back of the place and choose from menu options like the Long-EZ (two eggs your way), the White Knight (bacon and eggs), and Space-ShipOne (ham and eggs).

Mojave Airport established its outer space connection in 1990, when Orbital Sciences Corporation's Pegasus dropped from a B–52 mother ship and rocketed a communications satellite into orbit. The rocket's airframe, which included wings for pitching up from its horizontal drop-off attitude to reach space, had been built by Scaled Composites. Since the rocket was not reusable, Scaled continues to build Pegasus airframes for Orbital, which now launches them from an L–1011 jet that can frequently be spotted at Mojave Airport.

In 1996, space entrepreneur Gary Hudson founded the Rotary Rocket Company and set about trying to build the first private spaceship at Mojave Airport. As with Orbital's Pegasus, the spacecraft's airframe was built by Scaled Composites. Thirty million dollars and three years later, the project got off the ground, just barely, with a seventy-five-foot-high test flight of the helicopter-inspired Roton test vehicle flown down one of the runways by Brian Binnie and fellow Rotary test pilot Marti Sarigul Klijn. Rotary ran out of money and folded early in 2001, but the wheels (or was it helicopter blades?) had begun turning. The little town of Mojave, by then long known as a center of innovations in aviation, now began to be thought of as a place from which to get even higher. "It's not the edge of the world," people who live and work in Mojave have been known to say about their town, "but you can sure see it from here."

After Rotary folded, several of the company's engineers stayed on at the airport as XCOR Aerospace. They saw that the dream of private spaceflight was tantalizingly close to realization and that Mojave, with its edge-of-the-world status, miles of uninhabited desert all around, and rarely cloudy skies, made an ideal launch pad for rocket ships.

There's not a whole lot else to do in Mojave but build and pilot flying machines. Burt Rutan certainly didn't choose to live there for the scenery. "I remember being quoted in a magazine once saying it's a crummy little desert town," he told a group of local high school students in 2006, "and I've had to face up to that quote for a long time. But I have to say, I live in that crummy little desert town, and I think it's very special."

Call it what you will; just don't call Mojave dusty. So says Bill Deaver, a Mojave resident since 1948 and the editor and publisher of the *Mojave Desert News*. He's had just about enough of journalists from the East Coast breezing through the place, taking a quick look around, and reaching for the standard desert cliché about dusty streets. It's windy, sure, says Deaver, but Mojave just plain isn't dusty. With the phrase "dusty streets," Deaver hears journalists telling people in Mojave "We're from New York and you guys are a bunch of peons."

Like many of the residents of this area surrounded by Air Force bases (China Lake lies to the northeast of Mojave, on the other side of town from Edwards) and defense contractors that have traditionally depended on Republican politicians for defense spending, Deaver is a political conservative. I'm a vegetarian liberal agnostic from the Northeast—and not just any place in the Northeast, but that bastion of hippies and Vietnam-era peaceniks, Woodstock. On a visit to Mojave, I thought I'd be branded an interloper and run out of town on one of its many rails.

Instead, I got introduced to a Mojave Chamber meeting as a welcome guest from "the great state of New York." People loved the idea that I came from Woodstock to visit Mojave. "*That* Woodstock?" Yes, I admitted, just a bit apprehensively: *that* Woodstock, the one for which the famous rock festival of the summer of 1969 was named. Far from looking at me askance, however, the people I met in Mojave seemed delighted. It was as if they saw in me a fellow maverick.

That reaction surprised me at first—until Deaver pointed out to me over a steak dinner honoring outstanding high school students (the Edwards Air Force Base civilian employee sitting next to me happily accepted my steak for his doggie bag) that the extreme left and extreme right ends of the political spectrum have more in common with each other than with the middle. That was so, I agreed. And so it was with Mojave and Woodstock, I realized. At first glance utterly dissimilar, the two supercharged small towns had a lot more in common with each other than with the rest of the country.

Each was a hundred miles from the biggest city on its coast (Los Angeles and New York) and hence attracted people who appreciated all that the big cities had to offer but who also wanted to keep city life definitively at arm's length. Yes, both towns were peopled by mavericks—and not just any mavericks, but people dedicated to fundamentally shifting their perspectives beyond the routines other people grind through each day.

In Woodstock, that attempt at perspective shift tends toward an exploration of inner space: through meditation, music, the arts. Woodstockers get together to chant in Sanskrit, read their poems, and bang drums at the Village Green each Sunday afternoon. They study the Bhagavad Gita the way many middle Americans study their Bibles. They practice yoga and meditate at the yoga studios and Buddhist temples that dot the landscape like the churches of more typical small American towns. In the summer of 2006, Tibet's spiritual leader, the Dalai Lama himself, dropped into town during his world travels to deliver an address in one of the town's ball fields. Peace and love isn't just a cliché left over from the hippie era in Woodstock; it's a spiritual pursuit.

The people of Mojave are more likely to search for God from the cockpit of an airplane than through meditation. Mojave is the perfect place from which to fling themselves over the edge of ordinary reality and into the great blue beyond, and lately even farther than that: into the infinite black void of space, where, as Mike Melvill said after his first spaceflight in *SpaceShipOne*, "you really do feel you've reached out and touched the face of God." It seems a strange idea on the surface, but it's quite possible that the two towns are mirror images of the same American dream— the right of the individual to pursue his or her own bliss, to fly. In Woodstock that flight is more perceptual; in Mojave, more literal; but either way it's the will and the courage to do things differently, to find new worlds, whether through art or through engineering.

Burt Rutan is Mojave's resident guru, the grand master of aerospace arts, but there are many practitioners here. Cathy and Al Hansen, for instance. They came to the area around the same time Deaver did. They keep a hangar full of old war birds on the flight line—Huey helicopters; a couple of fighter jets; a Russian Antonov An–2 biplane, choice of bush pilots in Soviet Siberia; sports-car-sized minichoppers—all of which they rent out for movies and commercials to support their flying habit, repainting them depending on the job—mainly different schemes of military black and green.

Outside their hangar, the Hansens keep a more ordinary fleet of single-engine civilian airplanes, along with some military ground vehicles like a Scorpion Tank and a British FV–432 armored personnel carrier. For everyday getting around town, Cathy drives a Ford Explorer sporting bumper stickers that say "A Great American Hero: Ronald Reagan" and "I love the sound of jet engines." Active in the community, she writes a column for the *Mojave Desert News* and sits on the Mojave Airport board of directors. She radiates a youthful energy that belies her actual age. "I have one of those freedom-loving spirits, and I just love it out here," she says. "There's always something neat going on at the airport; there's no way to be depressed because you come out here and there's always something fabulous."

Just do it, is the Hansens' philosophy of life and the business of flying, and don't wait for someone to come along and tell you not to. That goes for the affairs of the airport itself. Al wasn't convinced it was such a good idea for airport general manager Stu Witt to ask for an $11 million state loan to build a passenger terminal and a hangar for Virgin Galactic. The money would come with strings; government money always does. Still, Cathy sees the increased government involvement in airport affairs as the inevitable price of success. "I think we need that," Cathy says of the new buildings. But, "I would like to see them not demolish the history that's here to make way for that, because I think there's room for both. It's important to know where you've come from to know where you are going."

Al remembers the old days, when if you wanted to put up a new building or fly a new airplane at the airport, you needn't seek anyone's approval. "All right, do it," Mojave's first civilian airport director Dan Sabovich told Al when he wanted to build on the flight line. "If you don't screw it up, you can do it." In fact, Sabovich expressed some irritation with Al for telling him about his project in advance, since that made him feel obligated to check it out, when he'd rather follow a "don't ask, don't tell" policy.

It's that kind of free spirit that the Hansens want to preserve in Mojave—a spirit they feel government involvement has a way of stifling. "I love the freedom so much," she tells me as the three of us stand in front of the couple's hangar doors. But with development encroaching on the area from Los Angeles to the south, and all the attention focused on Mojave's new status as a spaceport, "it's really fragile."

"It's always been a frontier town," Cathy says of Mojave, "and it still is, because we've gone from the railroad and gold mining to the frontier of space." Like any frontier town, Mojave "has always been a place for free-thinkers and very, very little regulation," agrees Al. "The things that are happening now have been happening over a long period; they change a little bit going from some certain kind of airplane to a rocket ship."

The Hansens do acknowledge that life on the frontier requires a certain tenaciousness. "You gotta be tough to live in Mojave," Cathy tells me. It's so damned hot in the summer, for one thing, though Al disputed that at once, saying it wasn't anything like Phoenix—only 110 degrees Fahrenheit instead of 120 plus. He did allow, though, that the wind made life in Mojave challenging. "We live in a river of wind," Cathy agreed, though it didn't sound so bad the way she put it. Not as if for three months of the year the wind blows at thirty knots every single day, sometimes gusting up to a hundred, and like as not to take the roof off your house.

Mojave Air- and Spaceport, officially known as the East Kern Airport District and Civilian Flight Test Center, is managed by an ex–fighter pilot in his mid-fifties named Stuart O. Witt. Everyone calls him Stu for short, but that apparently warm and fuzzy nickname belies a rattlesnake-tough personality only thinly veiled behind a cool smile.

Witt grew up on a ranch in Onyx, California, in the mountains a couple of hours' drive north of Mojave. After college he attended the Navy Fighter Weapons School, the famous TOP-GUN, and then flew F–14 Tomcats off an aircraft carrier. He was sixteen years old when Armstrong and Aldrin walked on the moon. He figured—along with just about every other red-blooded American boy—that he'd one day get his chance to leave the planet. Instead, says Witt now, "My contract has been broken by a lot of people who told me that in my lifetime I would have a chance to go to space. To date 440 people have made the trip and I'm not one of them."

That bugs him. A lot. "My generation went to the moon and then abandoned the moon," he says. "I'm trying to help create an environment that will allow all of us to actually realize a dream that many of us share." Specifically, he's working to turn Mojave into a booming spaceport with the same laser-beam intensity he brought to TOPGUN school and aircraft carrier flying.

Witt's office, when I visited the airport in the spring of 2006, seemed like nothing so much as a lair, a warrior's den. It was as big as the pilot's lounge down the hall, with room enough for a long leather couch and a coffee table in addition to a big wooden desk. On a side table near the couch, a collection of clubs and knives lay displayed on an animal skin. A heat lamp warmed a terrarium on a table nearer to Witt's desk. I didn't have the nerve to peer inside to see what creature dwelled within. Airplane models lined the windowsill behind Witt's desk. The window looked out on the flight line and the parked jetliners and the desert and Tehachapi Mountains beyond.

Witt was sharp-featured and fit. He moved with a fluid grace and appeared to measure each statement before he spoke, giving nothing away without due consideration to its strategic value. Although his manner seemed unhurried, almost lazy, I had the distinct impression of coiled potential lying just beneath the surface. I tried to resist the impulse to call him "Sir."

Witt had just been to the state capital to attend hearings and to press for his $11 million loan with which to build a pair of new buildings for Virgin Galactic. It wasn't much to ask to accommodate the business that could be the Next Big Thing, Witt thought. His goal was nothing less than to turn Mojave into the Silicon Valley of spaceflight, at the same time preserving the airport's primary mission to foster cutting-edge flight testing and research. To that end, he was working hard to attract as many NewSpace companies as he could to Mojave. To do that as effectively as possible, he felt he needed the support of the State of California.

Trouble was, California was host to a lot of Next Big Things.

Like Silicon Valley itself. And Hollywood. And a thriving aero-space industry powered by the U.S. Department of Defense and NASA. California legislators had a lot more to occupy them than a request for a several-million-dollar loan from an airport on the edge of the world, especially since, as Jennifer Gress, an analyst working for the Senate Transportation and Housing Committee put it, that money would only provide "millions for a billion-aire."

"In effect," said Gress in a report to the committee, if the loan bill passed, "the state would be subsidizing the construction of a building to house operations for a company whose owner has a net worth in the billions ($2.2 billion in 2004). It is unclear that these funds are needed to support Virgin Galactic. The commit-tee may wish to consider whether it is appropriate for the state to provide a loan that would support the operations of a private, for-profit enterprise."

For Witt, it wasn't just about the money; it was about be-ing able to say to prospective spaceport tenants that California wanted their business. "When you go interview for a job, you know the most powerful words you can say to somebody?" he asked me in our meeting. "'I. Want. The. Job.' You know how few people say that? I tell my staff when they interview—for years I've said that when you interview a professional—in the first sixty seconds, if they don't say 'I want the job,' it's an academic exercise; it's just a social event." The way Witt saw it, going to potential tenants of Mojave spaceport with real cash said "I want the job" as nothing else could. Sure, Branson was a billionaire. So what? If the state of New Mexico offered him hundreds of mil-lions of dollars to locate there and the state of California offered him squat, why shouldn't he interpret that as a hearty welcome from the one place ("I want the job!") and a cold shoulder from the other ("What's in it for me?").

That argument made a lot of sense to me. Airlines aren't ex-pected to build their own airports, and auto companies don't have to build highways. Yet, neither the airline nor the auto industries could survive without the necessary infrastructure funded with public money. And so it will be with the spacelines; to succeed, commercial spaceflight will have to rely on a similar pairing of private and public efforts. Branson and the other spaceline operators will have to locate where they are welcomed by local governments.

At least, Witt implored the senators of the Transportation and Housing Committee, even if they couldn't give him the loan he needed ("a *hand*, not a hand*out*," he said), "go on TV and say 'we want your business.'" No one had even taken that basic step, Witt told me.

In spite of Witt's best efforts to make the case for Mojave Spaceport in senate hearings, the loan bill, sponsored by state senator Roy Ashburn, got shot down three votes shy of the seven it needed to advance from the Transportation and Housing Committee to the full Senate. But Witt hadn't played his last card. He got on the phone. He called MSNBC's Alan Boyle, a space enthusiast and a commercial spaceflight booster. Boyle's report on the California legislature's rejection of the spaceport loan bill hit the Web as fast as Boyle could pound it out. In his story, Boyle reminded readers that the state of New Mexico was going all out to support its embryonic commercial spaceport, having just approved tens of millions of dollars free and clear for building new facilities. And he pointed out the simple irony that the only state that had actually hosted commercial spaceflights couldn't even be bothered to extend a measly $11 million loan to support the new industry.

Witt smiled one of his cool, almost predatory, smiles as he told me how, in his opinion, Boyle's article turned the crucial number of legislators around so that by the time his plane touched down in Mojave, the committee had voted again and this time had advanced the bill to the full Senate. The press, as far as Witt was concerned, wielded power.

"Do you want to see *SpaceShipTwo*?" he asked me.

Well, sure, of course. So did everybody. But Rutan was keeping that particular bird wrapped up tight and away from the prying eyes of the press, just as he had done while building the first spaceship. Surely Witt, of all people, knew that.

And yet.

He got on the phone with someone at Scaled, whose workshop was tantalizingly close to where I now sat, but so far out of any reporter's reach that it might as well be on the back side of the moon. Still, my heart quickened to the possibility: maybe Witt could get me in where nobody else could. I imagined myself climbing into Witt's car (it would be some gigantic American SUV, naturally) and taking the extremely brief drive to the Scaled hangar. I imagined the hangar doors rolling back just enough to let Stu—old buddy, old pal—and me in, where I'd be treated to sights Alan Boyle could only dream about.

"No? Okay, well, that's too bad." Witt was saying, or something to that effect. He hung up the phone.

He smiled, as if to say, well I did my best. Had he just been playing with me? Working to earn my trust?

I was standing before his desk by then. I suddenly felt like a subordinate summoned to the office of his superior, even though I was the one who had asked for this meeting. We exchanged some small talk. I stammered something out about how it had been an honor to see him. It was all too much for my poor little rabbit brain. Then Witt turned to his work, the universal sign for "It's time for you to go now."

One of the new space companies Witt had succeeded in attracting to Mojave was an outfit called Masten Space Systems. Masten Space was started in the San Francisco Bay area in 2004 by Dave Masten and a couple of fellow members of his local amateur rocket club, the Experimental Rocket Propulsion Society (ERPS). Masten was a mechanical engineer by training, who worked for computer networking giant Cisco Systems and had spent his life dreaming of space. Just a year older than I, he had also missed out on most of the excitement of the first space age.

Masten and his crew's idea was straightforward. For a hundred bucks, they would launch your soda-can-sized payload into space on an unmanned vertical-takeoff-vertical-landing rocket. They'd work their way up to orbit, and then start thinking about putting people into space. Unfortunately, unlike Elon Musk or Robert Bigelow, Dave Masten didn't have much money of his own. So he cobbled together what he could through small-ticket investors—enough to outfit a no-frills rocket shop in Santa Clara and begin building small lox/alcohol rocket motors to give himself and his engineers some experience.

Trouble was, the Bay Area, with its dense population, wasn't exactly ideal for testing rocket motors. So Masten and company outfitted a specially designed trailer that looked a lot like a hot dog stand. It had computers for gathering data; tanks for lox, alcohol, and helium (for pressurizing the propellants); and, pointed out the door, a mount for a rocket motor.

It was an elegant solution to a vexing problem, a sort of Murphy bed of rocketry. The Masten crew would tow the thing behind Masten's pickup into the hills behind Santa Clara to fire their rocket motors in one of the few suitably desolate areas they could find. Ultimately, though, it was a stopgap, one that limited them to motors small enough to put on the trailer. Not to mention the extreme hassle of towing their entire test rig, along with all their rocket fuel and lox and an electric generator, an hour out of town every time they wanted to fire a motor. This was no way to run a spaceship company. In June 2006, Masten Space Systems moved to Mojave Airport, with Stu Witt's blessing. I went to visit them in July, less than a month after the move.

Masten's new headquarters, across Sabovich Street from Scaled Composites, seemed little improved from the airport's Marine base days. With few interior walls, the low wooden building was unfinished except for a couple of smallish offices and a rest room. A spidery-looking rocket-powered flying machine squatted just inside the garage-style loading doors at the rear. About six feet high and slightly less than that around, it couldn't carry people, or even reach space, but it certainly looked otherworldly, with shock absorbers on its legs, pads for feet, and two big tanks in its complicated-looking core: one for lox, one for isopropyl alcohol. It was ringed with bolted-on bullet-shaped tanks for pressurizing the lox and alcohol tanks with a helium-nitrogen mix. It looked as if some space aliens had stashed their planet-hopper in an abandoned building while they went exploring on foot. Actually, Masten and his crew were building the machine as a test bed for another vehicle that would compete in the first annual Lunar Lander Challenge, a $1.35-million-dollar NASA-sponsored contest to be held at the X PRIZE Cup in Las Cruces, New Mexico, that October.

Adding to the feeling of desolation, in the back of the shop was an old broken-down car missing an engine. It had been towed all this way from Santa Clara by Masten Space engineer and automotive enthusiast Ian Moore, a rocketeer in his twenties with a mechanical engineering degree. Moore's girlfriend was "supposedly" coming out to join him in Mojave.

I had my doubts about anyone with mere blood in his or her veins, rather than rocket fuel, sticking it out in Mojave just to be with a loved one. Tough as the place could be to live in for those who couldn't pursue their passions anywhere else, it was an order of magnitude more difficult for the uninvolved spouses. In fact, I'd heard Burt Rutan joke earlier that year about the toll Mojave took on families who relocated to work there. "We used to keep a record of this," he said, "and that is: if you come from out of state to work at Scaled Composites, how long continuously does your wife cry after she sees Mojave? The record was seven and a half weeks."

The only woman on Dave Masten's crew that summer of 2006 was Arabella Mueller, a freshly graduated aeronautical engineer working as an intern before going back to school for her master's degree. She had gone to engineering school, hoping to work for NASA when she got out, and maybe even get to space that way. When she was two years into her studies, however, the flights of

Masten Space Systems test bed rocket squats in the company's workshop in Mojave, California. *(Courtesy of Masten Space Systems)*

SpaceShipOne convinced her that the future of space travel wasn't in government but in NewSpace.

I drove out with Mueller and Moore and Pierce Nichols, a mechanical engineer and software entrepreneur Masten had recruited from the ERPS, to the dusty patch of desert just north of the airport proper, where the Mojave rocket shops keep their test stands.

Yes, damn it, dusty; I saw an actual dust devil whirling across the hardscrabble, just picture perfect. It's all brown scrub brush out there as far as the eye can see, and dry as hell, so that a green plant there draws the eye like water or the blue infinity above.

Mueller and Moore got the hot-dog-stand test rig set up, charged it with helium from tanks chained to its side, and ran lines from the electric generator. They worked at the foot of a windowless two-story prefab building left by Rotary Rocket, but they didn't have access to it, so we all had to stay outside in the sun. Mueller and the boys filled the lox tank from a dewer (essentially a big pressure bottle for holding gases) on the back of Nichols' Toyota pickup. They set up a long folding table and some folding chairs on the other side of the prefab building from the test stand to keep us shielded from the rocket engine. On the table went a couple of laptops that would control the engine, run the test stand, and collect data.

Masten had driven up in his pickup, and the crew was just about ready to start test-firing one of the little five-hundred-pound-thrust rocket motors that would power Masten's Lunar Lander Challenge entry, when a bearded, rugged-looking man on in his forties interrupted the proceedings, wanting to know what we were doing there. Masten had just been telling me about his plans for the future, how his rockets would revolutionize access to space by enabling even high school classes to fly experiments that formerly were the province of major corporations and governments. He seemed every bit the space visionary then: in command, putting his dreams to metal and fire, out there with his crew working for the future good of humanity. But his spirits seemed to sag when confronted by the newcomer. He clammed up, letting Nichols do most of the talking.

The bearded man was a contractor for Gary Hudson's AirLaunch. He brought Masten and Nichols and me back to Hudson's test area across the road to discuss the situation. He had a field office there in a cargo shipping container; a fenced-off perimeter; a big, fixed test stand; and permanently installed fuel and lox tanks—the works.

Thing is, the contractor explained, you guys are squatting on land we're leasing. Our insurance won't cover you guys—or, more importantly, us—if your engine blows up. He showed us on a map taped to a wall inside the shipping container. Yes, Masten's portable rig was clearly within the boundaries of the leased land. Masten tried, not very hopefully, to explain that the airport had granted him permission to set up where he was, to no avail. The contractor showed Masten and Nichols on the map where they could move their test rig, just outside the boundary of the lease, and that was that. Moving the stuff would take a fair amount of time and effort, but Masten didn't protest. He just hiked back through the brush, muttering, "What other snag can we hit?"

I was feeling a little faint in the midday heat, and we were all getting pretty hungry by the time Moore and Mueller and Nichols got the hot dog test stand moved and the engine inside pointed at a relatively vegetation-free patch of desert. We then got behind Rotary Rocket's abandoned prefab, and Mueller and Pierce and I poked our heads out around the corner to watch what would happen when Moore, sitting at the impromptu command center at the folding table, hit the command to fire the motor.

A loud hiss. Puffs of escaping gas. But no light.

Masten and crew had succeeded in firing this particular motor at their old test site in the hills above Santa Clara, but the test stand evidently needed some tuning up after its five-hundred-mile road trip down the coast. The rocketeers tinkered, then tried again.

Again, the hiss and puffs of vapor, but nothing else.

Finally, on the third attempt, a bright orange flame blasted out of the test trailer, ruffling the head of a little green bush plant a hundred feet off, and damned if that tough little Mojave plant didn't take it. It stood right up to that flame as if it was just another one of those stiff Mojave breezes that cooled not a whit, just brought more heat.

Then, after only a second or two, the engine made a popping sound and sprouted flames around its base, burned through a bunch of wires there, and conked out just before Nichols hit the "scram" switch to vent all the lox and helium and keep the fire from spreading.

The crew broke for lunch back at the airport proper, and afterwards settled in for a long day working on the rocket engine and the test stand, prepared to spend as long as it took to get the rig running properly. You have to be tough to make it in Mojave, Cathy Hansen had told me, and now I saw firsthand exactly what that meant. Did Dave Masten have what it took? Did his company? No way to tell at that early stage, but I could see how they could go either way—fold, like Rotary Rocket, or struggle through until the cash started to flow and they could begin to breathe, like XCOR.

Masten and crew never did get a vehicle together in time for that year's Lunar Lander Challenge. But they redeemed themselves at the X PRIZE Cup with a crowd-pleasing series of perfect engine firings from their portable test stand. The display attracted the attention of several companies interested in hiring Masten Space Systems to help them with their own rocket projects, holding out the possibility for Masten Space to earn its first real income apart from investors. Masten's future, it seemed, was looking up.

Meanwhile, the state of New Mexico was throwing everything it had behind a new spaceport on undeveloped land in Upham, thirty miles to the east of Truth or Consequences. Support for the spaceport had been increasing for some time, but the effort to actually break ground kicked on the afterburners when Richard Branson held a press conference with Governor Bill Richardson and other state officials in December 2005. Together they announced that Virgin Galactic would make the spaceport its home base when it was completed, perhaps as soon as 2009. For Branson it was a no-brainer; after committing $120 million to developing and building his fleet of *SpaceShipTwo*s and mother ships, he wasn't about to pass up an opportunity to let someone else pick up the tab on a purpose-built spaceport, especially since much of it would be built to Virgin's specifications.

For New Mexico, the advantages weren't quite as clear cut. In fact it was a $225 million gamble that Virgin Galactic would not only get off the ground but prove to be a commercial success. Even more, that other commercial launch companies would follow. But the state wasn't operating completely in the dark. It had commissioned studies from space analyst firm Futron Corporation and New Mexico State University that concluded that the state had a lot to gain from a thriving spaceport. According to Futron's best estimates, New Mexico could expect to realize $460 million in annual economic activity from the spaceport by 2015. That money would come not only from space launches (more than two hundred a year by then, according to Futron), but also from increased spending by tourists visiting the area and by spectators flocking to events such as the X PRIZE Cup and Rocket Racing League races. The New Mexico State University study was equally bullish, concluding that the state could see $991.45 million to $1.2 billion in accumulated economic impact in the spaceport's fifth year of operation, which would include up to $357.21 million in earnings and support 2,871 jobs.

Then, too, the state wouldn't take on all of the estimated $225 million development costs for the spaceport. The state legislature approved the release of just $100 million in state money for the spaceport; with any luck, the rest would come from local and federal funds. The money would initially go toward putting roads into the area, and then runways, launch pads and other infrastructure, but the crown jewel would be a brand-new passenger terminal and other facilities for the spaceport's anchor tenant, Virgin Galactic.

True to Branson and Virgin's environmentalist slant, the facility would seek to have as little impact on the surrounding environment as possible. Most of it would be built underground, with only runways, a control tower, and a giant representation of Branson's eye, part of Virgin Galactic's new logo designed by renowned designer Phillip Starke, showing above ground. Passengers would board their spaceships in an underground hangar and then taxi out into the open for their takeoff runs. Power would come from solar panel arrays, and the facility would recycle the water it used. Branson was his usual jovial self at the press conference. "We may be even able to allow those aliens who landed here at Roswell fifty years ago in their UFO a chance to go home," he joked.

The state that had launched America's first suborbital rockets (Wernher von Braun's captured World War II—era V–2s, from the White Sands Proving Grounds, now White Stands Missile Range) wanted to carry its space legacy into the future. "It's absolutely fitting that our state, New Mexico, be the headquarters for outbound space travel, since many believe that Roswell, New Mexico, hosted in-bound space in 1947," said the governor at the press conference. "In fact," he joked, "some of the aliens are here today. I want to welcome them. They *are* here. They *are* in Roswell."

Along with Virgin Galactic, the Rocket Racing League planned to set up a base at the completed spaceport. In the meantime it cut a deal with the city of Las Cruces for some hangar space at the Las Cruces Airport, already the interim site of the annual X PRIZE Cup. In 2006, the spaceport changed its name from the prosaic Southwest Regional Spaceport to the grander-sounding Spaceport America, more closely reflecting its ambition to become the hub of commercial spaceflight in the United States.

Yet, it wasn't a tourist ship or a rocket racer that fired the spaceport's inaugural rocket, but rather a far less ambitious project by a little band of former amateur rocketeers called UP Aerospace.

UP Aerospace president Jerry Larson stands with one of his company's rockets at the 2005 X PRIZE Cup. *(Photo: Michael Belfiore)*

UP Aerospace's business plan called for launching solid-fu-eled rockets on a wild, largely uncontrolled, thirteen-minute ride into suborbital space. After arcing through its apogee as high as a hundred forty miles straight up, the little rocket—a solid-fuel-packed metal tube with stabilizing fins on its tail and an airframe and nose cone containing a payload and a simple track-ing transponder—would descend back through the atmosphere on parachutes for recovery by its ground crew. Capable of lift-ing only a hundred ten pounds, the sixteen- to twenty-foot-tall rockets would carry experiments for students and university and government researchers; small objects for memento seekers; and even the ashes of the deceased.

I met the UP Aerospace crew at the first X PRIZE Cup in the fall of 2005. Jerry Larson, the group's president, stood by himself next to his rocket while the spectators swarmed the more exciting-looking displays. As I approached I saw that the rocket looked very different from a lot of the others. "Yeah," Larson. "That's because it's *real*." And so it was. In an industry still more about hype than substance (though that was quickly changing), a piece of func-tional space hardware stood out among displays largely made up of mock-ups. I could tell that the rocket was made to fly, not just look cool. The aluminum body was finely machined, smooth and solid to the touch, without rivets or obvious welds.

In September 2006, UP Aerospace's concrete launch pad, command trailer, and fifty-six-foot-high launch rail were the only structures at Spaceport America. In the predawn hours of Sep-tember 25, reporters and spectators bumped over the still largely unimproved roads out to the site, escorted by police cars to show the way. Launch was scheduled for 7:30 a.m., but a problem with the rocket's transponder forced a delay while the ground crew took off the nose cone and fixed it. Finally, at just after two in the afternoon, Rick Homans, New Mexico's head of economic devel-opment, hit a switch to fire the rocket. It launched with a flash-boom-whoosh and rose on a fast-growing pillar of white smoke, ruler straight into the cloudless blue sky. Mission accomplished.

Or so it seemed until just nine seconds into the flight, when the contrail abruptly went jagged. The motor cut off thirteen seconds into the flight as planned. The rocket continued coasting to just 42,000 feet, far short of the 328,000 feet its builders had hoped it would reach, then pitched over and fell back to Earth. Days later, Larson and crew were still trying to find their rocket for an autopsy that would tell them what had gone wrong.

It seemed an inauspicious start to New Mexico's spaceport bid, and it highlighted the still-uncertain nature of rocketry, private or otherwise. To go truly mainstream, rocket launches would have to become a lot more predictable, especially when they had paying passengers on board. *SpaceShipOne* had provided a taste of what regularly scheduled spaceship launches could look like— not a single one of the ship's three spaceflights was delayed, and each one made its target altitude and returned safely. Now it was up to the industry as a whole to match that performance.

What would a future in which space travel was commonplace look like? It was no more possible to determine that for certain than it would have been for Charles Lindbergh to predict jumbo jets carrying five hundred people at a time across the Atlantic in just a few hours. But that shouldn't stop anyone from engaging in a little informed speculation.

10

The Sky's No Limit

Speculations on the Future

Let's go back to the future for a look at where the new space age might take us. It's 2034, thirty years after SpaceShipOne broke the government monopoly on space travel, and the 1960s dream of commonplace space travel has been a reality for some time now. Spaceports have grown up in empty lands around the world, creating a network of destinations for hyperspeed travel from point to point on the Earth. Although these spaceports are located safely away from big population centers, high-speed electric maglev lines now link most major cities with spaceports within a couple of hundred miles of them.

Although still mainly a research and development center for new types of air- and spacecraft, Mojave Spaceport is also now a major destination for people all over the globe, with suborbital and orbital flights arriving and departing daily. To get there, you can catch an air taxi from just about anywhere in the country to Palmdale Airport and get on the spaceport train there, or fly a bigger passenger jet into Los Angeles International Airport, the rail's originating station. LAX looks much the same as it did at the turn of the century: a big airport surrounded by the headquarters of major aerospace firms, only now SpaceX has taken its place among Boeing and Raytheon and the others with an office tower and sprawling fabrication facilities.

The train's doors whisper closed as you settle back for the thirty-minute ride (including stops) from LAX at three hundred miles per hour right down the center of the 405 and Antelope Valley freeways. Riding a magnetic pulse that alternately draws and repels it along, the train never actually touches the ground but levitates a few inches above its guideway to reduce speed-sapping friction.

Your train stops to pick up passengers in Palmdale, then heads out into the desert. Spaceport legislation has managed to halt the relentless march of development north from L.A., so that the buffer of largely uninhabited, and eminently crashworthy, desert remains around the town of Mojave. The desert whips past your window, and the distant mountains scroll by like a video game landscape—faster, it would seem to anyone used to travel by automobile, than they have any right to do. Between the stretch of desert beside the railroad tracks and the mountains, you catch a glimpse of a vast field of radio antennas. This is part of a several-thousand-acre receiving station for cheap abundant power beamed down from space. The sun's energy, converted into electricity by orbiting solar power plants, has replaced fossil fuels as the lifeblood of the world's economy, and it powers the maglev trains and plants turning out cheap hydrogen fuel extracted from ordinary water.

Mojave has grown in the past thirty years. There's the eighteen-hole golf course outside of town, right next to the Virgin resort hotel. Both Virgin Galactic and Scaled Composites have new business offices, R&D centers, and hangars at the air- and spaceport, as do XCOR and several other space-related companies. Tourist buses roll through on tours of the birthplace of commercial spaceflight and the Rutan Home, but the basic character of the town remains. The steel-wheeled freight trains roll on, the windmills keep spinning, the jumbo jets remain parked in the desert at the north end of town (though there are more of them now), and the place is still peopled by proudly independent seekers of the ultimate high. The place has prospered greatly in the past thirty years, becoming something of a boom town, though most of the engineers, pilots, and other professionals working at the spaceport still live in outlying communities like Tehachapi and Rosamond.

The train slows automatically at the spaceport terminal, the doors sigh open, and you step out with your bags into a typically busy Friday morning at the spaceport. There goes a group of Japanese business executives in jackets and ties on their way to the 9:10 Virgin flight to Dubai for some weekend partying, closely followed by a trio of NASA astronauts in bright orange survival suits, helmets under their arms and trailed by a pair of bloggers, a reporter from the *New York Times*, and a videographer. They're headed for a commercial Falcon 9 flight chartered by NASA to an orbiting Bigelow Aerospace research complex. Oh, and one of the bloggers is going with them into space. His suit is a more subdued shade of mauve. Those relaxed-looking men and women in black flight suits at the bar across the corridor? Bigelow Aerospace employees awaiting their flight to an orbital Hilton so they can start a fresh tour of duty.

And what's that next to the bar? It looks like a video arcade. Ensconced in black faux-leather seats like those in the spaceships on the flight line outside, gamers maneuver through moonscapes that shift, bump, and roll past them in wraparound holographic displays. But these are no ordinary video games. The players in those chairs are remote-controlling actual lunar rovers over the surface of the moon in real time. There are now hundreds of such rovers on the lunar surface, precursors to manned commercial lunar landers that will take sightseers there for real. Affordable booster rockets of the kind pioneered by SpaceX have made them, as well as hundreds of other robots probing the solar system, possible. Commercial rovers on the surface of Mars, working in conjunction with a constellation of mapping and communications satellites, have blazed the trail for a crew of privately funded astronauts now on a two-and-a-half-year journey (six months there, and a year and a half on the surface waiting for the optimum alignment of the planets for the trip back).

Meanwhile, the opening of manned spaceflight to private enterprise has created a worldwide economic boom that has no precedent in history. Suborbital spaceships flying around the world attract no more notice than the jumbo jets that still lumber across the sky. Orbital ships carrying sightseers, scientists, government astronauts of all nations, and, increasingly, construction workers, are only slightly less common. The suborbital ships take off from and land on runways at this spaceport many times a day, and orbital ships—both the air-launched and the vertical variety—take off from here at least once or twice daily as well.

The NASA astronauts and their mini-entourage have stopped in the middle of the passageway, momentarily blocking your way. "Anything to say to your colleagues on their way to Mars?" the guy from the *Times* asks an astronaut.

"We wish them the best of luck." The terse reply is accompanied by the briefest of professional smiles. It's clear the woman doesn't much want to talk about the dream job of her "colleagues," whose mission is more than half privately financed. NASA still doesn't accept money from commercial interests, so, with its second moon program down for the count, it's had to remain content with the same kind of round-and-round-the-Earth space station tours of duty that have been its mainstay for decades. The thrill of that kind of work has paled beside the private missions now leaving Earth orbit with increasing regularity. In fact, the only reason this particular NASA mission is getting this much media coverage is because of the blogger heading up with the NASA crew—the new reporters-in-space program is part of NASA's never-ending quest to try to drum up Congressional support for space research.

NASA got out of the spaceship development business long ago—the ships that had been intended to replace the space shuttles, and eventually go on to the moon and even one day to Mars, never materialized. Instead, after spending $5 billion of an estimated $100 billion commitment, NASA bowed to public pressure to instead fly its astronauts to Earth orbit on the Falcon 9, which made its maiden flight in 2012, just in time for the opening of Bigelow Aerospace's first commercial space station.

The Russians struck the next blow to NASA's second moon shot in 2015, when a modified Soyuz space capsule sent a Hollywood film director, a camera operator, and a professional cosmonaut to the moon in a commercial flyby mission to make a 3-D movie. Price tag for this mission brokered by Space Adventures, the same American company that sent the first space tourists to the International Space Station: $200 million, or $99 billion $800 million cheaper than *Apollo 2.0*. To be fair, the commercial mission didn't land on the surface, but it nevertheless showed the writing on the wall: there would be no way to justify spending tens of billions of dollars of taxpayers' money to duplicate a stunt of the past century when privately funded missions were doing it for well under a billion. It would be only a matter of time before the first commercial mission actually landed on the moon. Meanwhile, moviegoers thrilled to the lunar landscape whipping by only sixty-two miles below them in glorious 3-D. Space was *in*, as it hadn't been since the days of Apollo.

Ironically, the second moon race had been won by the forces of capitalism, just as the first one had been. Only this time it was the commercially oriented Russian space program that defeated NASA. It was not an unexpected development; after the collapse of the Soviet Union in 1991, the Russian Space Agency (RKA) had no choice but to turn to alternative sources of funding to keep flying. It started by selling advertising space on the sides of its booster rockets and shooting commercials on the *Mir* space station in the 1990s, and then flew the first paying passengers into space on Soyuz space capsules at $20 million a ticket, beginning in the 2000s. Crass as some at NASA thought such commercialization of a national space program was, it nevertheless provided a clear mission for RKA, and that was to serve its customers as safely and reliably as possible. It also gave RKA a steady source of funding that wasn't tied to political whim.

The Soyuz had originally been designed for the first moon race, and it wasn't much of a stretch to use it for the second. Not much more was required for the movie-making mission than the addition of larger view ports and camera mounts on the outside. A so-called Block DM booster docked with the Soyuz in Earth orbit to send it looping around the back of the moon for a gravity-assisted return to Earth.

The Chinese dealt the fatal blow to Apollo redux in 2017, a year before NASA's original deadline for the second series of moon shots, when two government taikonauts descended to the Sea of Tranquility in a lander designed with Russian help, hopped a short distance to the *Apollo 11* landing site, and gleefully snapped some shots of themselves standing in front of the *Eagle*'s landing stage. Amid some fearsome saber-rattling from Washington, the taikonauts stopped short of uprooting the United States flag.

Meanwhile, private enterprise rocketed ahead. The first commercial spaceships were strictly Earth-watching cruises to suborbital space, but they spurred the development of bigger, faster, and longer-ranging ships, until by 2020, riders bought tickets not just to see space and the curvature of the Earth as an end in itself but to get from place to place far faster than had ever before been possible. Business executives, world leaders, and the wealthy led the way until economies of scale drove down the price of a round-trip suborbital ticket to the point where the merely financially comfortable could afford it. And the price continued to drop. The fast package delivery company Federal Express got into the game as soon as it was feasible to do so. "Based on our research and investigation," FedEx CEO Fred Smith had said back in 1986, "hypersonic aircraft would be economically viable in our business; most importantly, our customers, more than any others, need this kind of improved transoceanic speed."

Same-day package delivery to anywhere in the world within an air-taxi's range of a spaceport was a valuable service indeed for those who just couldn't wait—recipients of organ transplants, for example. Operators of assembly lines awaiting replacement of critical parts, like special circuit boards available only overseas, were another. When a shut-down assembly line could cost a manufacturer $200,000 an hour, it made good sense to have a burned-out board sent by same-day suborbital flight—as long as that flight cost less than the shutdown. Precious stones were also among the relatively small, high-value cargoes for which same-day worldwide delivery made sense in the commercial space age. As a rule of thumb, according to a study on commercial space transportation services commissioned by NASA in 1994, the cargo industry deems anywhere from 3 percent to 6 percent of an item's value a reasonable price to pay for shipping it. And some items are truly priceless if delivered quickly enough. A human heart, for example, can survive outside the human body for only four hours.

As space travel became a greater part of ordinary people's lives, the overview effect experienced by the first space explorers and described by Frank White in his 1987 book of the same name took hold on a large scale. As much as the *Apollo 8* photograph of planet Earth seen in context of its neighborhood in space had affected the public consciousness, it couldn't compare with the power of experiencing that context personally. The feelings of X PRIZE title sponsor Anousheh Ansari, who became the first woman to pay her own way into space, were typical of the hundreds of thousands of civilian space travelers who followed her.

Ansari blasted off in a Soyuz space capsule in 2006 on a ten-day trip to the International Space Station, and like every astronaut, cosmonaut, and taikonaut before her, she saw the Earth as it truly is, without national boundaries. "From the side windows in the little cabins and the docking compartment, where I sleep," she wrote home to Earth, "you see the complete curvature of the Earth against the dark background of the universe. This view is actually my favorite because you see the "Whole" not the "Parts." I always like to see the big picture before deciding or worrying about the pieces. I wish the leaders of different nations could do the same and have a world vision first, before a specific vision for their country."

Getting that big picture brought humanity together as never before. For the first time, large numbers of nonprofessional astronauts, including those world leaders Ansari had wished could share her view, now saw for themselves just how small and fragile our planet really is. Going beyond suborbital space into orbit, as increasing numbers of people were doing to support new commercial development in space, was an even more powerful experience. "I had my sleeping bag next to this big window," Ansari told talk show host Oprah Winfrey after her return from space, "and I could see the Earth go by. . . . You watch the Earth peaceful without borders and you wonder—how could people ever do things to harm it?"

When you could circle the globe in ninety minutes, nations on the other side of the planet seemed as close as neighboring cities. War, environmental devastation, and all the other horrors threatening to pull humankind from its lofty heights now took place in the context of a greatly shrunken planet. True, it was mainly the rich and powerful who got to travel in space at first, but it was precisely those people who had the greatest influence on world affairs, and their new perspective moved them to consider the whole planet as well as their own interests. They might have left the planet representatives of one country or another, but they returned as citizens of Earth.

Many of Earth's problems stemmed from the way its citizens used energy. Earth needed something to replace polluting finite oil as the fuel for the world's economies, just as oil had replaced coal before it and coal had replaced wood. To meet the needs of not just the present but also the future, in which the economies of the world continued to grow, a new energy source had to be inexhaustible, or at least fully renewable. It also had to be environmentally clean, technologically feasible, and affordable. One possibility proposed in the wake of the first space age met all the requirements but one. It took the opening of the final frontier by private enterprise to at last make it affordable.

In 1971, Ralph Nansen was one of thousands of aerospace engineers coming off Project Apollo with great hopes for America's future in space. He'd gone to work on Apollo as a structural engineer for Boeing in 1961, eventually taking charge of the propellant tanks for the Saturn V's first stage. Even as the last of the moon walkers returned home from the ultimate camping trip in 1972, he and his Boeing colleagues were at work on ideas for taking space travel to the next level. The moon rockets were expensive throwaways. What was needed now, everyone agreed, was something more like an airliner than an artillery shell, something that could fly many missions instead of just one.

The space shuttle, its proponents declared, would make space travel truly routine and would open the way for all kinds of exciting new developments—big wheel-shaped space stations, slowly spinning to provide artificial gravity; permanent outposts on the moon, trips to Mars in ships built in Earth orbit. In short, the space shuttle was envisioned as a crucial piece of the infrastructure for establishing a more or less permanent human presence in space.

In early 1971, NASA gave space shuttle production contracts to North American Rockwell and McDonnell Douglas but hedged its bets by giving study contracts to a team made up of Boeing and Grumman Corporation, and to competitor Lockheed. At stake was the biggest aerospace prize since Apollo—and America's future in space. Nansen headed the Boeing side of things.

From the beginning, the shuttle's designers envisioned it with a reusable first stage, a so-called fly-back booster. The booster would do the heavy lifting of getting the shuttle off the pad and through the thickest part of the atmosphere. After expending its rocket fuel and releasing the shuttle to fire its own smaller rocket engines, the fly-back stage would spin up its jet engines, circle around, and land back at the spaceport. Many increasingly routine flights over a period of years, or even decades, would spread out the shuttle system's development and construction costs to bring the cost of each flight down over time.

Grumman engineers envisioned a fly-back booster that would be built around the Saturn V's first stage, the S–1C with its five F–1 rocket engines each delivering an earthshaking 1.5 million pounds of thrust. The shuttle itself would use the J–2 rocket engines that had powered the Saturn V's second stage. Although built for throwaway rockets, the engines had been designed for the many test firings needed for preflight testing and, with proper maintenance, should be good for hundreds of flights on reusable vehicles. The beauty of this plan, thought Nansen, was that the tremendously expensive development effort that had gone into the moon rockets would continue to pay dividends on America's next spaceship. NASA even had two brand-new Boeing-built S–1Cs left over from cancelled moon missions stored at the Michoud fabrication plant in New Orleans. All that remained was to ship them to Boeing's factories in Seattle and build an airplane frame around them. Piece of cake. Airplanes were what Boeing *did*; it wouldn't be much different from building any big new airplane. Just add a frame, some wings, a tail, and a cockpit to those S–1Cs, along with a dozen jet engines for flying it home after dropping off the shuttle, and it would be good to go in two or three years, tops.

Unfortunately, the politicians and the NASA brass had serious concerns about that design, most of them political. NASA, which had enjoyed carte blanche during the moon race, now was under financial pressure from a president who was reeling from the high cost of getting to the moon and waging the unpopular Vietnam War at the same time, and who did not want to spend any more money than absolutely necessary on manned space-flight. Build the shuttle with no budget increases, the Nixon administration commanded NASA. Nixon didn't concern himself with what it would cost to fly the thing; that would be the next guy's problem, or the problem of the guy after that. At the same time, the enormous bureaucracy that had build the moon ships refused to slim down, even in the face of a less ambitious space program.

Faced with Nixon's budget squeeze, NASA's leaders embraced a scheme that would save the shuttle by making it less expensive to develop but that would sacrifice much of the shuttle's utility as a reusable spaceship as well as its ability to operate affordably.

The trouble with the fly-back booster was that it required springing for a whole separate vehicle, with all the expensive development and flight testing that would entail. And then there was the senator from Utah. Frank Moss, chairman of the powerful (as far as the space program was concerned) Senate Science Committee, wanted to bring space shuttle business home to defense contractor Thiokol, which proposed building a pair of segmented solid fuel boosters for the shuttle's first stage.

Slap those solid fuel boosters onto the sides of a giant throw-away fuel tank—which, by the way, could be built at the Michoud plant in Louisiana, where the Saturn Vs had been built to keep that facility alive—and the shuttle could blast to orbit on a so-called stage-and-a-half design that would be much cheaper to develop than a fly-back booster. Inelegant, inherently less safe, and more costly in the long run, but eminently expedient from a political point of view.

Nansen didn't like it, and neither did his colleagues at Boeing and Grumman. They were engineers, not politicians, and they knew their design would be cheaper to operate than any design with components that would be thrown away during every flight. They kept submitting plans for fly-back boosters, but NASA kept the stage-and-a-half design in play during the process for selecting the final design. As Nansen saw it, in addition to all the other factors working against the fly-back booster, managers at the Marshall Space Flight Center in Huntsville worried that using existing engines for the shuttle would put them out of work, and the Manned Space Flight Center in Houston saw a piloted booster as a threat to the manned orbiter that was in its purview.

More than thirty years after the fact, Nansen still remembered the day in 1972 when the fly-back booster got the ax as the worst day of his life. After the cancellation, Nansen found himself reassigned to the singularly uninspiring task of managing program-cost analyses as part of Boeing's effort to find new roles for itself in the post-Apollo world. Those were hard times for Boeing. So many people got fired from the company's Seattle headquarters that a couple of real estate agents sprung for a billboard outside the airport that read "Will the last person leaving SEATTLE—turn out the lights."

Even though he knew that he should be thankful for even holding on to a job, Nansen nevertheless couldn't help feeling gloomy as he took a last walkthrough of his former aerospace engineering department one day in 1973. Stopping to look over the shoulder of one of the young aerospace engineers seated at his drafting table, Nansen leaned in for a closer look. The engineer, Dan Gregory, was working on some kind of spaceship. That much was clear. But it wasn't any spaceship Nansen had ever seen. His troubles momentarily forgotten, he asked him what it was.

"Oh, that's the Big Onion," Gregory replied. "I'm designing it to launch solar power satellites."

"You'd better start at the beginning," Nansen told him.

"Haven't you heard about Peter Glaser's idea of generating solar energy in space and then beaming it to the ground?"

Nansen hadn't. Gregory filled him in. As he explained, Nansen's black mood lifted. *This* was what Apollo had been leading up to. *This* was what NASA—the entire country, no, the whole world—had to pursue as the encore to humankind's greatest achievement. From that moment on, Peter Glaser's Big Idea became Nansen's purpose in life.

In 1968, Peter Glaser, a mechanical engineer at management consulting firm Arthur D. Little Company in Cambridge, Massachusetts, published a deceptively simple-looking, five-page paper in the journal *Science*. In it, he proposed that humankind meet its future energy needs with power from the sun. "The conversion of solar energy to usable power is the only alternative to nuclear power for the distant future," Glaser said in his paper, especially since "the control of fusion is still the physicist's dream."

But, said Glaser, ordinary solar arrays on the ground wouldn't do the job—not with the day/night cycle, bad weather, and the atmosphere itself interfering with proper reception of solar energy, not to mention the lack of room for installing arrays big enough to put a significant dent in the energy needs of a heavily industrialized country like the United States. No, the only way to really do solar power properly would be to put those solar cell arrays where they would get the full benefit of the sun's energy, unobscured by the atmosphere and out of the Earth's shadow.

Glaser figured that a solar cell array 11.5 miles in diameter orbiting the planet could power the entire northeastern United States. This solar power satellite would be parked in geostationary orbit, 22,300 miles up, where it would orbit at the same rate that the Earth turned, so that it would appear to hover over the same spot above the Earth and allow for an uninterrupted power supply.

Of course, getting that power down to Earth had to be a big part of Glaser's plan. A 22,300-mile-long wire wasn't going to work. Instead, Glaser proposed transmitting the power in the form of microwave energy—ordinary radio waves—down to terrestrial receiving stations, where they would be converted back into electricity for delivery on traditional power lines.

Radio waves had some important advantages over the filtered sunlight that terrestrial solar power arrays had to deal with. They cut right through clouds and atmospheric haze, for one thing. And they could deliver a more highly concentrated output, so that the area of the receiving station could be much smaller than a terrestrial solar array providing the same amount of power.

The wireless transmission of power had first been proposed in the early 1900s by Serb-American electrical engineer Nicola Tesla and had been demonstrated in 1964 by William Brown, an electrical engineer working for Raytheon under contract with the U.S. Air Force. By the time Glaser published his paper in 1968, all the technological elements were in place for meeting the world's energy needs with an inexhaustible supply of power from the sun—all, that is, except for an affordable means of launching the components of those 105-square-mile solar cell arrays and assembling them in orbit.

That's where Nansen came in. With an oil embargo on, the United States government needed ideas for freeing itself from foreign oil. Boeing got funding from NASA and the Department of Energy to put some serious engineering work into Glaser's idea. Nansen got himself put in charge of the project, and it gained traction as a possible part of a new United States energy policy that would phase out oil in favor of new sources of power. Nansen's team set as their goal a five-gigawatt power station, or five times the capacity of a typical nuclear power plant. That kind of power would require a solar cell array twenty square miles in area and a receiving station (or "rectenna," as Brown called the rectifying antennas he developed) thirty square miles in area.

Concerns about the possible dangers of a high-energy radio beam blasting down from outer space were addressed by researchers who built and operated an experimental power transmitter and receiver in the Mojave Desert with funding from Jet Propulsion Laboratory in 1976. The transmitter beamed thirty kilowatts of power to a rectenna a mile away with an efficiency of 82.5 percent and at an intensity at the receiver twice as great as that expected for a solar power satellite, with no perceptible damage to any wildlife in the area. In fact, the birds that regularly flew through the beam with no apparent discomfort ended up actually damaging the rectenna instead, reported one of the researchers, JPL's Richard M. Dickenson.

Solar power satellite inventor Peter Glaser dismissed fears that a space-based solar power beam might cook anything in its path, like a microwave oven. "I have a standing offer," he said in a 1978 interview in the *Washington Post*, " to provide the wine and the salad to anyone who promises to eat the duck that flies through the beam—cooked or not."

Put a fence around the rectenna to keep people out of the most energetic section of the beam, which dropped off sharply at the edges, and you'd have no problems at all keeping exposure to the beam well within acceptable limits. In fact, the rectenna would absorb 99 percent of that radio energy, which meant that if it were elevated like the power lines that would deliver the electricity generated by the setup, you could actually make use of the land beneath it for agriculture. Graze cattle under it. Or put the rectenna elements on the roof of a big greenhouse, as suggested by industrial engineer and artist John J. Olson.

After Boeing completed its solar power satellite study under Nansen's direction, the United States government came within a hair of releasing funds for a demonstration satellite. The House of Representatives passed a bill authorizing funds for the project, but it was abandoned by the Department of Energy in 1980 and never taken up in the Senate. Word from the White House was that President Carter would never sign off on such a bill, recalled Nansen, so its supporters couldn't drum up backing in the Senate.

The fact was that with the compromised space shuttle, the United States lacked the means to build a solar power satellite; a solar power satellite of the kind suggested by Nansen's study group at Boeing would require hundreds of launches to get the prefabricated components into orbit for assembly by construction crews. So it was that the solar power satellite idea, briefly a vital part of the public debate on United States energy policy, dropped out of sight.The Japanese government, with its paucity of oil reserves, took up the solar power satellite challenge in 1993 with a test in suborbital space during which a free-falling transmitter successfully beamed power in the form of radio waves to a "daughter" receiver. And yet, not even the best-laid plans of those who saw solar power from space as the solution to the ongoing energy crisis could change the basic reality of the situation: the affordable spaceships needed to make the idea feasible simply didn't exist.

In 2034, the solar power satellites, the "powersats," as they're known, are invisible during the day, but at night they come out stitched across the night sky, looking like a neat row of stars. They glow brightly in reflected sunlight high above, the twenty-first-century equivalent of the oil rigs that used to burn through the night along the coastlines of the oil-producing nations. Unlike the rigs, though, the powersats are there to stay—a hundred years and more, with routine maintenance to their station-keeping thrusters and the moving parts keeping their big, disk-shaped transmitters precisely pointed Earthward and their vast solar cell arrays always facing the sun like great silica flower petals.

Long-lived—and, of course, tapped into the cosmic connection of the limitless fusion dynamo in the sky that gives life to planet Earth—the powersats have allowed us to begin to realize our true potential as human beings at last. Now, for the first time in history, there's plenty of power to go around. The developing nations striving to achieve first-world living standards at last have the resources to achieve their dreams. All nations are freed of the growth-stunting effects of conservation and the need to war over finite energy reserves, and everyone can enjoy a world that has begun to heal from the polluting excesses of the past. With the advent of endless cheap electricity, automobiles, airplanes, and most spaceships now run on clean-burning hydrogen derived from ordinary water split into its elemental components of hydrogen and oxygen through the electrically powered process of electrolysis.

Not everyone agrees that solar power satellites are the answer to Earth's energy woes. "Space solar is ridiculous," says SpaceX chief Elon Musk. He believes that even with relatively affordable commercial rockets like his Falcon series, space solar power won't make economic sense. "In fact, I think if you instantly transport solar panels from Earth to orbit for free, it would still be more expensive than putting them on Earth." With improvements in technology, Musk sees terrestrial solar power generation as the key to Earth's energy future. He even founded a terrestrial solar power company called SolarCity to help things along, one household and business customer at a time.

Physicist Marty Hoffert of New York University is more enthusiastic. "SSP [space solar power] offers a truly sustainable, global-scale, and emission-free energy source," he says. But even he doesn't believe that private enterprise alone can muster the resources needed to make it work on a large scale. He says the government will have to do the job, starting with a small demonstration satellite. He also advocates using lasers instead of microwaves to beam the energy down from space. With their more concentrated focus, lasers will allow for a much less massive infrastructure, though their inability to penetrate cloud cover will limit them in other ways.

Still, whatever form the ultimate space moneymaker takes, it will involve developing a new frontier, and frontiers have always given people a reason to hope and to dream—itself a worthy goal. And that's the real legacy of a hand-built minivan-sized craft called *SpaceShipOne*.

In the fall of 2005, the Smithsonian Institution acknowledged the historical importance of the little spaceship that could by installing it in the National Air and Space Museum on the

National Mall in Washington, D.C. There, the museum's nine million annual visitors could view it hanging in a place of honor in the museum's Milestones of Flight gallery, right beside those other epoch-making flying machines *The Spirit of St. Louis*, the Wright Flyer, and the *Apollo 11* command module. As was only fitting, for "*sic itur ad astra*"—thus one goes to the stars!

WHERE ARE THEY NOW?

New developments in the field of commercial spaceflight now occur almost daily, making it nearly impossible for any one person to keep track of them all. That's a problem I love to have, and that's why this is a book about beginnings rather than an attempt to survey all that flies on rocket power and private cash. Nevertheless, here's an update on some of the people in this book as we went to press.

After leaving Scaled Composites, Tim Pickens, the rocketeer who inspired Burt Rutan to use hybrid propulsion for SpaceShipOne, founded Orion Propulsion in Madison, Alabama, where he builds rocket motors and runs test stands for both private and government rocketeers. His philosophy: "Sell shovels to the miners." He wants to provide the means for others to build spaceships without trying to build one himself.

Scaled Composites was on schedule to roll out *SpaceShipTwo* for public view by the end of 2007.

Jim Benson left SpaceDev, *SpaceShipOne*'s propulsion contractor, to found another company called Benson Space Company, which intends to purchase Dream Chaser spaceships built by his former company for suborbital flights. One of his first hires was former space shuttle astronaut Robert ("Hoot") Gibson as chief pilot.

John Carmack was sure his Armadillo Aerospace could have launched its hydrogen-peroxide-powered X PRIZE vehicle (though not to space and not with the prize weight) before the prize deadline—if only his fuel supplies hadn't dried up. Instead, the team had to start over with new lox-alcohol engines, which the team publicly demonstrated for the first time at the 2005 X PRIZE Cup. Armadillo's *Pixel* rocket was the only Lunar Lander Challenge competitor at the next year's Cup, but the squat, unmanned machine tipped over on landing, ending its bid to win the prize money. The team still planned to reach space in the near future.

Also at the 2006 X PRIZE Cup, XCOR and the Rocket Racing League unveiled the first X-Racer.

Brian Feeney abandoned his balloon-launched X PRIZE vehicle concept and founded a new company called DreamSpace to build a liquid-fuel-rocket-and-jet-powered suborbital spaceplane.

Masten Space Systems successfully test-fired its rocket engines attached to the Lunar Lander Challenge test bed vehicle in its own test area in Mojave. The team looked forward to competing against Armadillo at the 2007 Lunar Lander Challenge.

UP Aerospace determined that not having enough tail fin surface area, not a flaw in the propulsion system or electronics, had sent its Spaceport America–opening rocket off course. It added a fourth tail fin to its rocket design and was preparing for its second launch attempt from Spaceport America.

New Mexico governor Bill Richardson announced his candidacy for president.

Elon Musk's SpaceX launched its Falcon 1 design for the second time in March 2007 from Kwajalein. The rocket climbed to 200 miles in altitude before a roll problem in the second stage prevented the dummy satellite on board from reaching orbital velocity. Still, Musk said that SpaceX had "flight proven 95-plus percent of the Falcon 1 systems" and that it was on track to launch a satellite for the Department of Defense in late summer.

Bigelow Aerospace loaded its second small-scale space station test module, *Genesis II*, with small objects and photos, including an early design for the jacket of this book, and was preparing for its launch from Russia.

Notes

Prologue: Full Circle

6   "To see the earth as it truly is": Archibald MacLeish, "A Reflection: Riders on Earth Together, Brothers in Eternal Cold," *New York Times*, December 25, 1968, p. 1.

Space or Bust

13   Diamandis and Maryniak's airplane ride: Gregg Maryniak, telephone interviews, February 6 and February 9, 2006.

15   Details of Diamandis' inspiration in *The Spirit of St. Louis* and his quotes: Peter Diamandis, telephone interview, February 14, 2006.

15   "As a stimulus to courageous aviators": David Nevin and the editors of Time-Life Books, *The Pathfinders*, Time-Life Books, 1980, revised 1985, p. 51.

15   Lindbergh's competitors: Nevin, pp. 51–107.

16   " . . . behaving as though Lindbergh walked on water . . ." the number of people who turned out to see Lindbergh in person: A. Scott Berg, *Lindbergh*, Berkley Books, 1998, p. 170.

16   Lindbergh's effect on the aviation industry and *Forbes* quote: Berg (pp. 171), except for the increase in air travelers, which is from Roger E. Bilstein, *Flight in America: From the Wrights to the Astronauts*, The John Hopkins University Press, 1984, 1994, and 2001, p. 57.

16   The term "Lindbergh boom": Smithsonian Institution National Air and Space Museum Web site, "Milestones of Flight, Ryan NYP 'Spirit of St. Louis,'" http://www.nasm.si.edu/galleries/GAL100/stlouis.html (accessed February 22, 2005).

17   Diamandis' life and career history: Effie Lascarides, *Apollo's Leg-*

acy: *The Hellenic Torch in America at the Dawn of the New Millennium*, Hellenic College Press, 2000, pp. 62–79.

18 " . . . my 'mission' in life . . .": Lascarides, *Apollo's Legacy*, p. 64.

19 Diamandis' conversations with Lichtenberg and his decision to go to space without NASA's help: Peter Diamandis, telephone interview, February 14, 2006.

20 Clarke's axiom: Arthur C. Clarke, *Report on Planet Three and Other Speculations*, Harper & Row, 1972, p. 70.

20 " . . . an example of a person who dramatically changed the way people thought about aviation with a small-scale effort": Gregg Maryniak, "The High Frontier Vision: 1993 Status and Strategy," *Space Manufacturing 9: The High Frontier: Ascension, Development and Utilization: Proceedings of the Eleventh SSI-Princeton Conference*, American Institute of Aeronautics and Astronautics, 1993, p. 381.

20 "Ten million bucks is chicken feed": Gregg Maryniak, telephone interview, February 9, 2006.

21 History of aviation prizes: Gregg Maryniak, "When Will We See a Golden Age of Spaceflight?" *Space: The Free-Market Frontier*, Cato Institute, 2003; Henry Serrano Villard, *Contact! The Story of the Early Birds*, Bonanza, 1968.

23 Space tourism studies: P. Collins, R. Stockmans, and M. Maita, "Demand for Space Tourism in America and Japan, and Its Implications for Future Space Activities," Space Future, http://www.spacefuture.com/archive/demand_for_space_tourism_in_america_and_japan.shtml (accessed February 26, 2006).

23 "We have met the payload of the future, and it's us!": Gregg Maryniak, telephone interview, February 9, 2006.

23 X PRIZE rules: X PRIZE Foundation, "Ansari X PRIZE Competition Launches Commemorative Program," 2004, p. 1.

25 Connectors concept and quotes: Malcolm Gladwell, *The Tipping Point: How Little Things Can Make a Big Difference*, Little, Brown and Company, 2000, pp. 38, 45–46.

25 Students for the Exploration and Development of Space, ISU, and associated quote: Lascarides, *Apollo's Legacy*, p. 69.

26 " . . . if Kerth liked your idea, you were halfway there" and associated details, including text of Kerth's letter and quote from Korte: Eli Kintisch, "Founder's Bold Dream Soars Toward Reality," *St. Louis Post-Dispatch*, September 22, 2004, p. A1.

26 March 4, 1996: X PRIZE Foundation, "History of the X PRIZE," http://www.xprizefoundation.com/prizes/xprize_history.asp (accessed February 27, 2006).

26   Andy Taylor details and quote: Joan Dames, "St. Louisans Catch Out-of-World Spirit, *St. Louis Post-Dispatch*, May 26, 1996, p. 2.

27   Erik Lindbergh's lunch with Diamandis and Lichtenberg: Erik Lindbergh, telephone interview, February 24, 2006.

27   "I see things in wood": Erik R. Lindbergh, Lindbergh Gallery Web site, http://www.lindberghgallery.com/default.asp?lk=misState (accessed March 6, 2006).

27   "Can't we use 10 million bucks better here on Earth?" and associated Lindbergh quotes: Erik Lindbergh, telephone interview, February 24, 2006.

28   O'Neill's space vision: Gerard K. O'Neill, *The High Frontier*, Bantam, 1976, 1977. Quote on p. 60.

29   "Everything that we hold of value on this planet": Peter Diamandis, telephone interview, February 14, 2006.

30   "Martian smoke trails blasting out of rustic rocket ships": Erik R. Lindbergh, Lindbergh Gallery Web site, http://www.lindberghgallery.com/default.asp?lk=misState (accessed March 6, 2006).

30   Lindbergh's transformation and associated quotes: Erik Lindbergh, telephone interview, February 24, 2006.

30   Lindbergh's 2002 flight: "Lindbergh Repeats Epic Solo Flight," CNN Web site, http://archives.cnn.com/2002/US/05/02/lindbergh.france/index.html, May 2, 2002 (accessed March 3, 2006).

30   X PRIZE press conference and launch dinner: Eli Kintisch, "Founder's Bold Dream Soars Toward Reality," *St. Louis Post-Dispatch*, September 22, 2004, p. A1; X PRIZE Foundation, "History of the X PRIZE," http://www.xprizefoundation.com/prizes/xprize_history.asp (accessed February 27, 2006); *Overview: 1996 Gala Dinner (Burt Rutan speech), History, Announcement, Gala Dinner Event, The Future*, X PRIZE Foundation video provided by the X PRIZE Foundation.

31   "I want to win this thing": *Overview: 1996 Gala Dinner (Burt Rutan speech)*.

Go!

32   "How about this one?" I gathered my firsthand observations of Feeney and his da Vinci Project during the week of October 15–22, 2003.

34   Jim Akkerman was working on a suborbital spaceship: Most details of Akkerman's work and background come from Jim Akkerman, telephone interview, October 2, 2003.

34   A devout Christian, Akkerman had prayed for help: Eli Kintisch,

"Dreams-Turned-Schemes Launch One Spaceworthy Rocket Ship," *St. Louis Post-Dispatch*, September 23, 2004, p. A1.

34 Akkerman's explanation of "Advent:" Jim Akkerman, e-mail correspondence, January 8, 2007.

36 ASRM axed in 1993 amid cost overruns: "Implications of the Termination of the Advanced Solid Rocket Motor (ASRM) Program" abstract, Penny Hill Press Congressional Research Service Documents, http://www.pennyhill.com/index.php?lastcat=62&catname=Space+Activities&viewdoc=93–965 (accessed February 3, 2007).

36 Solid-fueled rockets as a bad choice for the shuttle, along with the reasons why: Mike Mullane, *Riding Rockets: The Outrageous Tales of a Space Shuttle Astronaut*, Scribner, 2006, pp. 32–33.

36 A two-stage rocket would bob upright in the water: Mark Carreau, "Rocket Ride Entrepreneurs Plot Barnstorming, 21st Century Style," *Houston Chronicle*, February 2, 1997, p. A1.

37 "I'm building this rocket myself": Kintisch, "Dreams-Turned-Schemes Launch One Spaceworthy Rocket Ship."

42 Avro Arrow: "Avro CF–105 Arrow," FlightDeck, a Web site of Discovery Channel Canada and the Canadian Aviation Museum, http://exn.ca/FlightDeck/Aircraft/Hangar2.cfm?StoryName=Avro%20CF%2D105%20Arrow (accessed April 9, 2006).

45 Carmack's pre-X PRIZE history: David Kushner, Masters of Doom: How Two Guys Created an Empire and Transformed Pop Culture, Random House, 2003.

45 Binary space partitioning: Kushner, *Masters of Doom*, p. 142; Carl Shimer, "Binary Space Partition Trees: Presentation for CS563, Advanced Topics in Computer Graphics," Worchester Polytechnic Institute Department of Computer Science Web site, http://web.cs.wpi.edu/☐matt/courses/cs563/talks/bsp/bsp.html (accessed April 2, 2006).

45 "1,000-horsepower monsters": John Carmack, telephone interview, September 29, 2003.

45 Ferraris faster than 200 miles per hour: John Carmack, e-mail correspondence, January 10, 2007.

45 A dozen iterations of *Doom* and *Quake* in 2000: http://www.idsoftware.com (accessed April 2, 2006).

46 A new skill just once every few months: John Carmack, telephone interview, September 29, 2003.

46 Exchanging e-mail to find a team to build "vertical dragsters" and the donation of Bob Norwood's shop: Kushner, *Masters of Doom*, p. 289; John Carmack, e-mail correspondence, January 10, 2007.

46   First members of Armadillo Aerospace, working Tuesday nights and Saturdays: Preston Lerner, "A Few Dreamers Building Rockets in Workshops," *Popular Science*, May 2003, p. 56.

46   "Soaking up huge amounts of information" and the other quotes in this paragraph: John Carmack, telephone interview, September 29, 2003.

47   Details of Armadillo's rockets, testing, and X PRIZE preparations: "Armadillo Aerospace News Archive," Armadillo Aerospace Web site, http://www.armadilloaerospace.com/n.x/Armadillo/Home/News   (accessed April 9, 2006).

THE HOMEBUILT SPACESHIP

55   The epiphany: Allison Gatlin, "They're Not Your Average Couple: Burt, Tonya Rutan Met on Blind Date," *Antelope Valley Press*, September 29, 2004, p. A1; Tonya Rutan, telephone interview, May 2, 2006.

56   Following that thought right off the golf course: Brian Binnie, interview in Rosamond, California, February 3, 2005.

56   Rutan brothers' childhoods and early careers: Vera Foster Rollo, *Burt Rutan: Reinventing the Airplane*, Maryland Historical Press, 1991; Jeana Yeager and Dick Rutan with Phil Patton, *Voyager*, Knopf, 1987.

57   "I had even more fun at home developing my own airplane": *Black Sky: The Race for Space*, Discovery Channel documentary, first aired September 28, 2004.

58   VariViggen details: Rollo, *Burt Rutan: Reinventing the Airplane*, pp. 19–21, 28; Yeager, Rutan, and Patton, *Voyager*, pp. 41–44; "Rutan VariEze," Web site of the Smithsonian National Air and Space Museum, http://www.nasm.si.edu/research/aero/aircraft/rutan_eze.htm (accessed June 20, 2006).

58   " . . . a real 'macho machine' . . . : "Rutan VariEze," Web site of the Smithsonian National Air and Space Museum.

59   "There was no question in my mind which I wanted to keep": Rollo, *Burt Rutan: Reinventing the Airplane*, p. 21.

59   The place had started life as a training base for Navy and Marine pilots: Mojave Chamber of Commerce brochure, "Mojave: California's Golden Crossroads," obtained June 2006.

59   Sold to home builders all over the country starting in 1976: "Rutan VariEze," Web site of the Smithsonian National Air and Space Museum.

60   Stopped designing new airplanes for the private market: Yeager, Rutan, and Patton, *Voyager*, p. 44.

60   "I spent hours drooling over the smooth, contoured, efficient glass

composite European sailplanes; Rutan's fiberglass-and-foam building technique: Rollo, *Burt Rutan: Reinventing the Airplane*, p. 39.

61   Like building a boat: Black Sky: The Race for Space.

61   "This method is light, strong, requires no particular skills or tools:" Rollo, *Burt Rutan: Reinventing the Airplane*, p. 40

61   *Popular Science* cover story in 1978: Ben Kocivar, "Hot Canard: Is This Foam and Fiberglass Home-Built the Shape of the Future?" *Popular Science*, November 1978, p. 84.

62   Details of RAF in the late 1970s: Mike Melvill, telephone interview, June 7, 2006.

63   The Fairchild contract; the beginning of Scaled Composites: Mike Melvill, telephone interview, June 7, 2006.

64   X–15 details and Rutan's thoughts about them: Dennis R. Jenkins and Tony R. Landis, *Hypersonic: The Story of the North American X–15*, Specialty Press, 2003; Burt Rutan, telephone interview, June 4, 2006.

65   X–15 accident: Jenkins and Landis, *Hypersonic*, pp. 149–152, 196–198; Milton O. Thompson, *At the Edge of Space: The X–15 Flight Program*, Smithsonian Books, 1992, pp. 261, 263; Mike Melvill, telephone interview, June 7, 2006.

67   Rutan's early spaceship sketches: Cory Bird, telephone interview, June 22, 2006.

67   Details of Rutan's three-person capsule: Burt Rutan, talk given at Experimental Aircraft Association annual fly-in convention, Oshkosh, Wisconsin, 1997, from a videotape provided by the X PRIZE Foundation; Cory Bird, telephone interview, June 22, 2006.

69   Early feather designs and foam models: Cory Bird, telephone interview, June 22, 2006; *Black Sky: The Race for Space*; Burt Rutan, e-mail correspondence, February 13, 2007.

70   Dethermalizer: Dan Kreigh, telephone interview, June 15, 2006.

70   Bird and Kreigh looked at each other: Cory Bird, telephone interview, June 22, 2006.

71   Details of Pickens's early life: Tim Pickens, interviews in Huntsville, Alabama, March 24, 2006.

72   "I look around Huntsville and wonder what happened to all the smarts:" Tim Pickens, e-mail correspondence, December 6, 2000.

73   "We just concluded that we needed to take my bicycle and one of my engines" and other quotes; details of the rocket bike: Tim Pickens, telephone interview, October 22, 2005; my own observations of the rocket bike in Las Cruces, New Mexico, October 6, 2005.

73   "You'll find a lot of guys in this business who theorize to death . . .:" Tim Pickens, telephone interview, June 29, 2006.

74   RP–1: "Countdown! NASA Launch Vehicles and Facilities," Section 2, Propellants, NASA Facts Online, John F. Kennedy Space Center, http://www-pao.ksc.nasa.gov/kscpao/nasafact/count2.htm (accessed July 6, 2006).

75   Rutan hadn't designed propulsion for his previous craft and associated quote: Burt Rutan, telephone interview, April 6, 2006.

76   Details of Rutan's visit to Huntsville: Tim Pickens, telephone interviews, October 22, 2005, June 29, 2006, and March 3, 2007; in-person interviews in Huntsville, Alabama, March 24, 2006, and Las Cruces, New Mexico, October 6, 2005.

78   The dinner at Telini's additional details: Burt Rutan, e-mail correspondence, February 13, 2007; Tim Pickens, e-mail correspondence, February 16, 2007, and March 4, 2007.

SpaceShipOne, Government Zero

80   Exchange between *White Knight* and air traffic control: *Black Sky: The Race for Space*, Discovery Channel documentary, first aired September 28, 2004.

80   Like a Klingon warship: Peter Pae, "Private Spaceflight Is a Public Success," *Los Angeles Times*, June 22, 2004, p. A1

80   *White Knight* technical details and first two flight dates: "Posterboard—White Knight," Scaled Composites Web site, http://www.scaled.com/projects/tierone/data_sheets/html/white_knight.htm (accessed July 14, 2006); "White Knight Flight Test Summaries," Scaled Composites Web site, http://www.scaled.com/projects/tierone/logs-WK.htm (accessed February 3, 2007).

81   *White Knight*'s problems on first flight, emergency landing, and subsequent perfect flight: Brian Binnie, e-mail correspondence, February 2, 2007; Pete Siebold, telephone interview, January 30, 2007.

81   *White Knight*'s fiberglass construction: Cory Bird, telephone interview, June 22, 2006

81   Paul Allen's involvement and motivations: *Black Sky: The Race for Space;* press conference I attended at Scaled Composites, June 20, 2004.

82   Scaled Composites press conference in April 2003: Alison Gatlin, interview in Palmdale, California, April 24, 2006; *Black Sky: The Race for Space*; Rutan quotes are from *Black Sky: The Race for Space*.

82   Stars and stripes on *SpaceShipOne* and *White Knight*: Dan Kreigh, telephone interviews, June 5 and 15, 2006.

83   "Those guys are having too much fun!" Carl Hoffman, "The Right Stuff," *Wired*, July 2003, p. 135.

83   Hole-in-one insurance policy: Eli Kintisch, "Dreams-Turned-Schemes Launch One Spaceworthy Rocket Ship," *St. Louis Post-Dispatch*, September 23, 2004, p. A1; Alan Boyle, "SpaceShipOne Wins $10 Million X PRIZE," MSNBC, October 5, 2004, online at http://www.msnbc.msn.com/id/6167761/ (accessed August 3, 2006); Robin Snelson," Unsung Heroes of the Personal Spaceflight Revolution," *The Space Review*, September 27, 2004, http://www.thespacereview.com/article/234/1 (accessed February 2, 2007).

84   "The most scary flight": Mike Melvill, telephone interview, June 7, 2006.

84   Mike and Sally Melvill's early years with Rutan: Mike Melvill, telephone interview, June 7, 2006; Sally Melvill, telephone interview, June 8, 2006.

85   *SpaceShipOne* first glide flight details: *Black Sky: The Race for Space*; Brian Binnie, telephone interviews, August 19 and 29, 2005.

87   Tim Pickens's time with Scaled Composites: Tim Pickens, interview in Huntsville, Alabama, March 24, 2006; telephone interview June 29, 2006.

89   SpaceDev and eAc test firings: *Black Sky: The Race for Space*; Burt Rutan, telephone interview, April 6, 2006.

89   "They were very close": Burt Rutan, telephone interview, April 6, 2006.

89   Details of SpaceDev and eAc's contributions to *SpaceShipOne*: Scaled Composites press release, September 18, 2003, online at http://scaled.com/projects/tierone/091803.htm (accessed March 4, 2007).

89   Six more glide flights: "Tier One Private Manned Space Program: combined *White Knight/SpaceShipOne*'s Flight Tests," Scaled Composites Web site, http://www.scaled.com/projects/tierone/logs-WK-SS1.htm (accessed August 3, 2006).

89   15,000 pound thrust rocket motor: Tim Pickens, telephone interview, June 29, 2006.

89   Details of *SpaceShipOne*'s first powered flight: "Tier One Private Manned Space Program: Combined *White Knight/SpaceShipOne* Flight Tests," Scaled Composites Web site, http://www.scaled.com/projects/tierone/logs-WK-SS1.htm (accessed August 3, 2006); *Black Sky: The Race for Space*, Brian Binnie, telephone interviews, August 19 and 29, 2005; Eric Adams, "The New Right Stuff," *Popular Science*, November 2004, p. 60.

91   Details of Pete Siebold's next powered flight the following April: Pete Siebold, telephone interviews, June 23, 2006, and January 30, 2007.

92   Melvill's powered flight in May: Black Sky: *The Race for Space*; "Tier One Private Manned Space Program: Combined *White Knight/SpaceShipOne* Flight Tests," Scaled Composites Web site, http://www.scaled.com/projects/tierone/logs-WK-SS1.htm (accessed August 3, 2006).

93   20,000 people: Derek Webber, "The Future Starts Here," *The Space Review*, June 28, 2004, online at http://www.thespacereview.com/article/173/1 (accessed August 3, 2006).

93   "I was absolutely in a state of shock . . .": This quote and others, along with details of the flight: post-flight press conference in Mojave, June 21, 2004, from my recording; "Tier One Private Manned Space Program: Combined *White Knight/SpaceShipOne* Flight Tests," Scaled Composites Web site, http://www.scaled.com/projects/tierone/logs-WK-SS1.htm (accessed August 3, 2006).

93   First time Rutan had invited public and press to a test flight: Burt Rutan, e-mail correspondence, February 13, 2007.

94   "This is not good": Black Sky: *The Race for Space*; Mike Melvill, telephone interview, June 7, 2006; Burt Rutan, e-mail correspondence, February 13, 2007.

94   Details of what the stuck trim stab meant, Melvill's reactions, and the banging noises: Black Sky: *The Race for Space*; Mike Melvill, telephone interview, June 7, 2006; Pete Siebold, telephone interview, January 30, 2007.

97   Siebold's cancer scare and decision to give up X1: Pete Siebold, telephone interviews, June 23, 2006 and January 30, 2007.

99   Tormenting himself: Brian Binnie, "Confessions of a Spaceship Pilot," *Air & Space Smithsonian*, June/July 2005, p. 28.

99   Tighe's assurances and first quote: Black Sky: *The Race for Space*.

99   Viscous dampers: Black Sky: *The Race for Space*; Jim Tighe, telephone interview, June 8, 2006.

99   "Brian did a phenomenal job": Jim Tighe, telephone interview, June 8, 2006.

99   Viscous dampers fix: Brian Binnie, e-mail correspondence, March 8, 2007.

99   Binnie's childhood and aviation career: "Brian Binnie," Web site of Maxwell-Gunter AFB, http://www.au.af.mil/au/goe/eaglebios/05bios/binnie05.htm (accessed October 28, 2005); Karen Bale, "First Scot in Space," *Daily Record* (Glasgow), October 6, 2004, pp. 23–4; Kathy Kiely, "Rocket Man," *Princeton Alumni Weekly*, February 23, 2005, online at http://www.

princeton.edu/%7Epaw/archive_new/PAW04–05/09–0223/features4.
html (accessed October 28, 2005).

101 Mike Melvill quotes and details of practice glide flights, telephone interview, June 7, 2006.

101 Binnie's time at Rotary Rocket: Marti Sarigul-Klijn, telephone interview, June 20, 2005.

102 $190,000 million ticket fees: "Virgin Group Sign Deal with Paul G. Allen's Mojave Aerospace," Virgin Group press release, September 27, 2004.

103 Most of Binnie's quotes and observations as well as most of the technical details of flying *SpaceShipOne* come from Brian Binnie in telephone interviews on August 19 and 29, 2005, and an in-person interview in Rosamond, CA, February 3, 2005.

104 "Numb toes . . .": Brian Binnie, "Flying for the X Prize," *Plane & Pilot*, June 2005, p. 60.

104 Radio exchanges between *SpaceShipOne,* mission control, *White Knight,* and chase plane: AOL X PRIZE Experience on the Web at http://channelevents.aol.com/research/xprize/index.adp (accessed March 10, 2005).

105 Melvill didn't see it that way: Melvill's perception of events and associated quotes come from Mike Melvill, telephone interview, June 7, 2006.

108 "He was at 213,000 feet . . .": Scaled Composites Tier One flight log at http://scaled.com/projects/tierone/logs-WK-SS1.htm (accessed August 25, 2005.

108 " . . . a realm of blessed peace and quiet": Brian Binnie, "Flying for the X Prize."

108 " . . . vast presence, looming and yawning . . .": Brian Binnie, "Flying for the X Prize."

109 Interactions between Rutan, Branson, and Allen: AOL X PRIZE Experience on the Web at http://channelevents.aol.com/research/xprize/index.adp (accessed March 10, 2005).

109 The 354,200-foot mark set in 1963: Dennis R. Jenkins and Tony R. Landis, *Hypersonic: The Story of the North American X–15*, Specialty Press, 2003, pp. 226, 228.

110 The top of its arc: 367,500 feet: Scaled Composites Tier One flight log at http://scaled.com/projects/tierone/logs-WK-SS1.htm, accessed August 25, 2005).

111 "Keep your head down, and swing smooth": From my recording of the press conference.

NASA Hitches a Ride

112 Text of H.R.5382, the Commercial Space Launch Amendments Act of 2004 (Enrolled as Agreed to or Passed by Both House and Senate): THOMAS Legislative Information on the Internet, http://thomas.loc.gov/home/thomas.html (accessed September 12, 2005).

113 "Tombstone mentality" and "Experimentation with human lives" quotes: Jeff Foust, "Oberstar Strikes Back," *Space Politics*, February 10, 2005, http://www.spacepolitics.com/archives/000435.html (accessed January 18, 2007).

113 "Strangled in its crib . . ." and " . . . Daredevils and visionaries. . . .": "House Floor Debate on Commercial Space Launch Amendments Act of 2004 (House of Representatives—November 19, 2004)," SpaceRef.com, http://www.spaceref.com/news/viewsr.html?pid=14565 (accessed September 12, 2005).

114 "Someone *will* die": Peter Diamandis, speaking on a panel at the International Space Development conference, Arlington, Virginia, May 21, 2005, from my recording.

114 "The 30,000 people who've now spoken to us about wanting to fly . . ." and worries about the commercial space industry's future: Will Whitehorn, speaking at the International Space Development Conference, Arlington, Virginia, May 21, 2005, from my recording.

116 Text of President Bush's moon speech: "President Bush Announces New Vision for Space Exploration Program," White House press release, January 14, 2004, online at http://www.whitehouse.gov/news/releases/2004/01/20040114-3.html (accessed December 14, 2006).

116 Chinese manned space program and Oberg quote: "Testimony of James Oberg: Senate Science, Technology, and Space Hearing: International Space Exploration Program," April 27, 2004, SpaceRef.com, http://www.spaceref.com/news/viewsr.html?pid=12687 (accessed December 14, 2006).

117 Kennedy's moon speech: "Special Message to the Congress on Urgent National Needs," delivered May 25, 1961, Web site of the John F. Kennedy Presidential Library and Museum, http://www.jfklibrary.org/Historical+Resources/Archives/Reference+Desk/Speeches/JFK/003POF03NationalNeeds05251961.htm (accessed December 14, 2006).

117 NASA's initial moon ship design: "NASA Releases Plans for Next Generation Spacecraft," NASA press release, September 19, 2005, online at http://www.nasa.gov/home/hqnews/2005/sep/HQ_05266_ESAS_Release.html (accessed December 17, 2006).

118 Only $1 billion a year for space shuttle development and ensuing design compromises: Dennis R. Jenkins, *Space Shuttle: The History of Developing the National Space Transportation System*, second edition, self-published, 1992–1997, p. 107.

118 The genesis of t/Space: Gary Hudson, telephone interview, July 30, 2005.

119 The first TV commercial shot on the International Space Station: "RadioShack First To Air TV Spot Filmed on the International Space Station—Commercial Launches May 27 for Father's Day," RadioShack press release, May 25, 2001, online at SpaceRef.com, http://www.spaceref.com/news/viewpr.html?pid=4973 (accessed December 14, 2006).

119 David Gump's lunar rover plan: Rex Ridenoure and Kevin Polk, "Private, Commercial and Student-oriented Low-cost Deep-space Missions: A Global Survey of Activity," paper presented at the 3rd International Academy of Astronautics (IAA) International Conference on Low-Cost Planetary Missions, April 27–May 1, 1998, Pasadena, CA, p. 6, online at http://www.smad.com/analysis/IAApaper-finaldoc.pdf (accessed November 20, 2006); Mary Motta, "Lunacorp and RadioShack Team Up to Send Robots to the Moon," Space.com, June 15, 2000, http://www.space.com/businesstechnology/business/lunacorp_moon_000615.html (accessed December 15, 2006).

119 Gump's conversations with Admiral Steidle: David Gump, telephone interview, December 12, 2006.

120 Air Force/DARPA FALCON program: "FALCON Force Application and Launch from CONUS Broad Agency Announcement (BAA) Phase I Proposer Information Pamphlet (PIP) for BAA Solicitation 03–35," July 29, 2003, online at http://www.darpa.mil/TTO/falcon/FALCON_PIP_FINAL.pdf (accessed December 15, 2006).

122 "We cut t/Space loose . . .": Michael Lembeck, telephone interview, June 21, 2005.

122 Jim Voss's thoughts on space shuttle seats and CXV seat development: Jim Voss, telephone interview, June 22, 2005.

122 Jim Voss career history: "James S. Voss," official NASA astronaut biography: http://www.jsc.nasa.gov/Bios/htmlbios/voss-ji.html (accessed December 15, 2006).

123 t/Space's proposal to NASA: David Gump, panel discussion and interview at the International Space Development Conference, Arlington, Virginia, May 19, 2005.

124 Lembeck's observations of t/Space booster drop test: Michael Lembeck, telephone interview, June 21, 2005; technical details of t/LAD and

test, Marti Sarigul-Klijn and David Gump, telephone interviews, June 2005.

126 "I was astounded that you could get this booster to do a 90-degree turn . . .": Chuck Coleman, telephone interview, June 30, 2005.

127 "I know how easy it is to throw out a set of viewgraph presentations . . .": Michael Lembeck, telephone interview, June 21, 2005.

127 Griffin's plan: "NASA Administrator Mike Griffin Remarks to the Space Transportation Association, Washington, D.C.—Tuesday, June 21, 2005," transcript by Constellation Services International, online at http://www.nasa.gov/pdf/119275main_Griffin_STA_21_June_2005.pdf (accessed September 12, 2005).

129 On a clear Wednesday morning . . .: David Gump, telephone interview, August 7, 2005.

130 Launch of COTS: "NASA Seeks Proposals for Crew and Cargo Transportation to Orbit," NASA press release, January 19, 2006, online at http://www.nasa.gov/home/hqnews/2006/jan/HQ_06029_Crew_Cargo_RFP.html (accessed December 17, 2006); about twenty-four applicants, six COTS finalists: NASA made no official statement of these numbers, but this information leaked from anonymous sources; Michael Belfiore, "NASA Makes First Round of Cuts for COTS," *Dispatches from the Final Frontier*, May 9, 2006, http://michaelbelfiore.com/blog/2006/05/nasa-makes-first-round-of-cuts-for.html (accessed December 17, 2006); Alan Boyle, "Finalists Picked in NASA's Private Space Race," MSNBC.com, May 10, 2006, http://www.msnbc.msn.com/id/12706352/ (accessed December 17, 2006).

130 Two companies funded for COTS: "NASA Selects Crew and Cargo Transportation to Orbit Partners," NASA press release, August 18, 2006, http://www.nasa.gov/home/hqnews/2006/aug/HQ_06295_COTS_phase_1.html (accessed December 17, 2006).

130 Hudson's additional financing: AirLaunch press releases, "Air Force C–17 Successfully Drops Prototype of Low-Cost Rocket," October 5, 2005; "AirLaunch LLC Completes Another Falcon Milestone," March 15, 2006; both online at http://www.airlaunchllc.com/News.htm (accessed December 17, 2006).

130 Gump continues t/Space: David Gump, telephone interview, December 12, 2006.

130 Elon Musk and SpaceX: see notes for "Orbit on a Shoestring."

131 Kistler Aerospace history: Greg Klerkx, *Lost in Space: The Fall of NASA and the Dream of a New Space Age*, Vintage Books, 2004, 2005, pp. 115–121.

131 K–1 75 percent complete; purchase by Rocketplane: Lon L. Rains, "Rocketplane's Majority Owner Buys Kistler," *Space.com*, February 27, 2006, http://www.space.com/news/rocketplane_022606.html (accessed December 17, 2006).

131 Kistler hardware in warehouses: Jeff Foust, "COTS Winners Start Showing Their Hands," *The Space Review*, October 30, 2006, http://www.thespacereview.com/article/733/1 (accessed December 17, 2006); Rocketplane Kistler business manager Chuck Lauer, interviews in Oklahoma City, Oklahoma, April 12–13, 2006.

132 Origins of the Rocket Racing League: Granger Whitelaw, telephone interview, November 9, 2005; Peter Diamandis, telephone interview, November 10, 2005.

132 Rocket racing the twenty-first century's big new sport: "Diamandis Launches Rocket Racing League: Premier Competition of 21st Century Unveiled," Rocket Racing League press release, October 3, 2005, online at http://www.rocketracingleague.com/media/press_releases/20051003_diamandis-launches-rrl.html (accessed December 17, 2006).

133 EZ-Rocket: I watched the EZ-Rocket in action both during a test run prior to demos at the first annual X PRIZE Cup in Las Cruces, New Mexico, on October 7, 2005, and during the two demos themselves on October 9, 2005.

133 EZ-Rocket technical details: "The EZ-Rocket: A Manned, Flying Rocket Test Vehicle," Web site of XCOR Aerospace, http://www.xcor.com/products/vehicles/ez-rocket.html (accessed December 17, 2006).

134 "Descend into management": Rick Searfoss, telephone interview, September 20, 2005.

134 Rick Searfoss career history: "Richard A. Searfoss," official NASA astronaut biography, http://www.jsc.nasa.gov/Bios/htmlbios/searfoss.html (access January 18, 2007).

137 "We were afraid we might have a law of physics problem," lox-fill technical details: Jeff Greason, telephone interview, November 15, 2005.

137 "They said, 'Amazing news!'": Peter Diamandis, telephone interview, November 10, 2005.

137 Advances in automobile technology from race cars to consumers: Rick Barrett, "Driving Innovation; from Seat Belts to Disc Brakes to Rear-View," *Milwaukee Journal Sentinel*, June 3, 2006, online at http://www.findarticles.com/p/articles/mi_qn4196/is_20060603/ai_n16456623 (accessed December 17, 2006).

138 "If you sat down to build a car . . .": Jeff Greason, telephone interview, November 15, 2005.

138 "You can clearly identify with the risks that auto-racing drivers take . . .": Fred Nation, telephone interview, November 23, 2005.

139 X-Racer top speed: "The EZ-Rocket: A Manned, Flying Rocket Test Vehicle," Web site of XCOR Aerospace, http://www.xcor.com/products/vehicles/ez-rocket.html (accessed December 17, 2006).

139 X-Racer engine burn times: Rick Searfoss, e-mail communication, November 28, 2005.

139 "I see innovation cycles coming on top of each other": Burt Rutan, speech at the Antelope Valley Board of Trade, February 25, 2005 (from my recording).

The 200-g Roller Coaster

140 "I used to get beat up, down, and sideways" and Chuck Lauer's beginnings in space development: Chuck Lauer, interview by David Livingston on *The Space Show*, December 3, 2006, online at http://www.thespaceshow.com/detail.asp?q=633 (accessed February 26, 2007).

141 Sushi bar in Oklahoma City, Oklahoma: This conversation took place on May 13, 2006.

142 Zero-g wedding dress: Dennis Overbye, "On the Runway: Spacewear Meant to Dazzle, Even in Zero Gravity," *New York Times*, May 16, 2006, p. F1.

143 Anderson's *Titanic* voyage: "Member Spotlight: Reda Anderson: Explorer . . . Adventurer . . . Phenomenon . . . Space Rider!" *Jonathan Magazine*, April 2006, p. 6.

143 Anderson's contract and Lauer's reaction to it: Chuck Lauer on *The Space Show*, December 3, 2006.

144 Virgin Galactic launch: "Virgin Group Sign Deal with Paul G. Allen's Mojave Aerospace; Licensing the Technology To Develop The World's First Commercial Space Tourism Operator," Virgin Group press release, September 27, 2004.

144 Branson's early years: Spencer Reiss, "Rocket Man," *Wired*, January 2005, p. 136.

145 Origins of Virgin Atlantic: "In the Beginning," Virgin Atlantic Web site, http://www.virgin-atlantic.com/en/us/allaboutus/ourstory/history.jsp (accessed August 23, 2006).

145 Branson's adventures: "Who's Richard Branson?" Virgin.com, http://www.virgin.com/aboutvirgin/allaboutvirgin/whosrichardbranson/default.asp (accessed August 23, 2006).

145 Virgin Galactic origins and Will Whitehorn quote: Will Whitehorn, telephone interview, January 23, 2007.

146 Rutan played hard to get: Burt Rutan, telephone interview, April 6, 2006.

146 New Mexico spaceport and details: Virgin Galactic/New Mexico press conference, Governor Bill Richardson, Richard Branson, Virgin Galactic president Will Whitehorn, and New Mexico secretary of economic development Rick Homans, Santa Fe, New Mexico, December 14, 2005, from my recording.

146 *SpaceShipTwo*'s passenger capacity, common fuselage with *White Knight 2*: "SS2, White Knight 2 to Use Common Fuselage," Web site of *Flight International*, July 3, 2006, http://www.flightglobal.com/Articles/2006/03/07/Navigation/200/205290/SS2%2c+White+Knight+2+to+use+common+fuselage.html (accessed August 23, 2006).

146 *White Knight 2* to be called *Eve*, wingspan of Boeing 737, passengers aboard *Eve* to watch space launches, and number of *SpaceShipTwos* to be built for Virgin: " SpaceShipTwo Mothership Based on Globalflyer," Web site of *Flight International*, August 16, 2005, http://www.flightglobal.com/Articles/2005/08/16/Navigation/177/200972/SpaceShipTwo+mothership+based+on+Globalflyer.html (accessed August 23, 2006).

146 Unrestrained flight aboard *SpaceShipTwo*: Leonard David, "Burt Rutan on Civilian Spaceflight, Breakthroughs, and Inside SpaceShipTwo," Space.com, August 11, 2006, http://www.space.com/news/060811_rutan_interview.html (accessed August 23, 2006).

146 *SpaceShipTwo* flight profile: Burt Rutan, speech to high school students, Mariah Country Inn & Suites, Mojave, California, April 27, 2006, from my recording.

148 Burt Rutan's plans to fly on *SpaceShipTwo*: Jeff Foust, "Virgin Galactic and the Future of Commercial Spaceflight, " *adAstra*, republished on Space.com, May 23, 2005, http://www.space.com/adastra/050523_virgin_nss.html (accessed August 23, 2006).

148 Victoria Principal and Bryan Singer on *SpaceShipTwo*: Associated Press, "Virgin Galactic Aims to Fly Passengers by 2008," July 18, 2006, online at http://www.space.com/news/ap_060718_virgin_update.html.

148 $11 million loan: Alan Boyle, "Spaceport Turnaround," Cosmic Log, MSNBC.com, April 19, 2006, http://www.msnbc.msn.com/id/12359455/ (accessed August 24, 2006); Leonard David, "California Lawmakers Back Mojave Spaceport Growth," Space.com, April 21, 2006, http://www.space.com/news/060421_mojave_cash.html.

148 " . . . not just the kind of crap you can see from New Mexico": Burt Rutan, speech to high school students, Mojave, California, April 27, 2006, from my recording.

149 Details of Reda Anderson's interactions with the engineers at Rocketplane Kistler and technical details of the *Rocketplane XP* come from a visit Anderson and I made to the company at the same time, April 12 and 13, 2006.

152 "Basically we're all whores for the ride" and other quotes: Mitchell Burnside Clapp, telephone interview, September 29, 2003.

152 Black Horse: Robert Zubrin, *Entering Space: Creating a Spacefaring Civilization*, Jeremy P. Tarcher/Putnam, 1999, pp. 45–49.

153 DC-X price tag, other details: Greg Klerkx, *Lost in Space: The Fall of NASA and the Dream of a New Space Age*, Vintage Books, 2005, pp. 103–110.

153 DC-X size compared with full-size spaceship and other details: "The DC-X and DC-XA Pages," NASA History Division Web site, http://www.hq.nasa.gov/office/pao/History/x–33/dcx_menu.htm (accessed August 24, 2006).

153 Fate of DC-X, details of X–33 and VentureStar: Klerkx, *Lost in Space*, pp. 99, 109; Zubrin, *Entering Space*, pp. 32–33.

154 Rocketplane's early years and the origins of the Rocketplane vehicle: Zubrin, *Entering Space,* p. 49.

155 "We thought if we could get one company to come to Oklahoma . . . ," other quotes, and details of tax credits: Gilmer Capps and Chuck Lauer, interview in Oklahoma City, April 12, 2006; Scott Cooper and Ben Fenwick, "Space Cowboys," *Oklahoma Gazette*, July 13, 2005, online at http://www.altweeklies.com/alternative/AltWeeklies/Story?oid=oid%3A148324 (accessed August 25, 2006).

155 "Citing creative differences . . .": Mitchell Burnside Clapp, telephone interview, May 30, 2006.

156 Urie's quotes and his background: David Urie, interviews in Oklahoma City, April 12 and 13, 2006.

157 "There's a big cost to designing the details of a fuselage": Bob Seto, telephone interview, May 17, 2006.

157 "It's a stout, robust airplane" and Learjet origins: Mike Klemanovic, telephone interview, May 12, 2006.

158 Oklahoma Spaceport at Burns Flat statistics and mission statement: Oklahoma Space Industry Development Authority fact sheet.

159 "It gives me chills right now thinking about it:" Reda Anderson, talk to elementary school students at Burns Flat, Oklahoma, April 13, 2006.

161 "The performance numbers given by the company are reasonable . . .": Dan Erwin, telephone interview, May 18, 2006.

162 Polaris Propulsion and details of the AR–36: Dave Crisalli and George Garboden, interview in Oxnard, California, May 2, 2006.

162 14,000-pound-thrust booster: David E. Crisalli, "700 Miles North . . . and 50 Miles Up?", *High Power Rocketry*, November 1997, p. 10.

164 Blue Origin: Sandi Doughton, "Amazon CEO Gives Us Peek into Space Plans," *Seattle Times*, January 14, 2005, online at http://seattletimes.nwsource.com/html/businesstechnology/2002150764_blueorigin14m.html (accessed December 17, 2006); Alan Boyle, "Blue's Rocket Clues," *Cosmic Log*, MSNBC.com, June 24, 2006, online at http://cosmiclog.msnbc.msn.com/archive/2006/06/24/669.aspx (accessed February 26, 2007); Michael Graczyk, Associated Press, "Private Space Firm Launches 1st Test Rocket," *Fort Worth Star Telegram*, November 14, 2006, p. B6.

165 " . . . hundreds of thousands of people flying outside the atmosphere soon": Burt Rutan, telephone interview, April 6, 2006.

165 "We think that the future for suborbital is really in point to point": Chuck Lauer on *The Space Show*, December 3, 2006, online at http://www.thespaceshow.com/detail.asp?q=633.

Orbit on a Shoestring

166 I made my observations at SpaceX's McGregor, Texas test facility and interviewed Tom Mueller, Tim Buzza, and others on January 14, 2005.

167 "You woke my baby last night": Tom Mueller, interview in McGregor, Texas, January 14, 2005.

168 Falcon 1's payload capacity: Elon Musk, telephone interview, December 10, 2004.

169 Established satellite launch prices: most launch providers keep their launch prices secret, making a direct comparison to SpaceX's published prices difficult. These estimates come from Jeff Foust, analyst at space business research firm Futron Corporation, telephone interview, October 2, 2006.

169 $6.7 million for a Falcon 1 launch: "Historic SpaceX Launch Set for December 19: The World's Lowest Cost Rocket to Orbit," SpaceX press release, December 15, 2005.

169 $400,000 airline tickets: Purchase price of a Boeing 737–600 is $47 million to $55 million (http://www.boeing.com/commercial/prices/index.html; accessed October 2, 2006), and a single-class configuration carries 132 passengers (http://www.boeing.com/history/boeing/737.html; accessed February 26, 2007). $55 million divided by 132 equals $416,666.

170 "The reason I started SpaceX" and other quotes in this paragraph:

Elon Musk, interview in El Segundo, California, May 2, 2006.

170 "... transportation is such a fundamental input...": Elon Musk, telephone interview, December 9, 2004.

170 Musk's early years, first businesses, and origins of SpaceX: Todd Halvorson, "Elon Musk Unveiled," *Florida Today*, January 29, 2005; Jennifer Reingold, "Hondas in Space," *Fast Company*, February 2005, p. 74; Seth Lubove, "Way Out There," *Forbes*, May 2003; Leslie Wayne, "A Bold Plan to Go Where Men Have Gone Before," *New York Times*, February 5, 2006; Brad Lemley, "Shooting the Moon," *Discover*, September 2005, p. 28; quotes and further details: Elon Musk, telephone interview, January 16, 2007; Elon Musk e-mail correspondence, February 1, 2007.

171 Zip2's offerings: "Virtual Media Partnerships," Zip2 Web site, December 2, 1998, available at the Internet Archive at http://web.archive.org/web/19981202110718/www.zip2.com/sites/About_Us_Zip2/html/our_company.html (accessed October 3, 2006).

172 "It's just that those who have built and operated them...": Brad Lemley, "Shooting the Moon," *Discover*, September 2005.

173 "If we ultimately wish not to go in the direction of the dinosaurs...": Elon Musk, telephone interview, December 10, 2004.

173 Reaction Research Society: "RRS Courses," Reaction Research Society Web site, http://www.rrs.org/RRS_Courses/rrs_courses.html (accessed October 3, 2006).

173 "Can you build something bigger?" Tom Mueller, interview in McGregor, Texas, January 14, 2005.

174 Musk in early 2005: I made these observations on a visit to SpaceX headquarters in El Segundo, California, on February 9, 2005.

175 "Ford didn't invent the internal combustion engine" and other quotes: Elon Musk, telephone interview, December 10, 2004.

176 Delta II configuration: "Delta II: Overview," Boeing Web site, http://www.boeing.com/defense-space/space/delta/delta2/delta2.htm (accessed October 3, 2006).

176 Pegasus configuration: "Pegasus User's Guide," Orbital Sciences Corporation, online at http://www.orbital.com/NewsInfo/Publications/peg-user-guide.pdf (accessed October 3, 2006).

176 Merlin design features: Tom Mueller, interview in McGregor, Texas, January 14, 2005.

177 Pintle injectors compared with other designs: Jeff Foust, "TRW Unveils New Low-Cost Rocket Engine," Space.com, September 26, 2000, http://www.space.com/businesstechnology/technology/trw_rocketengine_000926.html (accessed October 3, 2006).

177   Gas generator rocket engines: "Fastrac Engine—A Boost for Low-cost Space Launch," NASA Web site, http://www.nasa.gov/centers/marshall/news/background/facts/fastrac.html (accessed October 4, 2006); "Power Cycles," Aerospace Corporation Web site, http://www.aero.org/publications/crosslink/winter2004/03_sidebar3.html (accessed October 4, 2006).

177   Merlin's turbopump: "Rocket Engine Turbopumps," Web site of Merlin turbopump manufacturer Barber Nichols, http://www.barber-nichols.com/products/rocket_engine_turbopumps/ (accessed October 4, 2006).

177   Brian Bjelde and avionics details: Brian Bjelde, interview in El Segundo, California, February 9, 2005.

177   Falcon 1 diameter: David Darling, "Falcon 1 (rocket)," The Encyclopedia of Astrobiology, Astronomy and Spaceflight, http://www.daviddarling.info/encyclopedia/F/Falcon_1.html (accessed October 5, 2006).

179   *Challenger* disaster and shuttle range safety system: *Report of the Presidential Commission on the Space Shuttle Challenger Accident*, "Chapter III: The Accident;" "Chapter IX: Other Safety Considerations," June 6, 1986, online at http://history.nasa.gov/rogersrep/51lcover.htm (accessed October 5, 2006).

191   Vandenberg test run: Justin Ray, "Privately-made Falcon 1 Rocket Roars on the Pad," Spaceflight Now, May 27, 2005, http://spaceflightnow.com/falcon/050527frf/ (accessed October 1, 2006).

192   The move to Kwajalein: Leonard David, "SpaceX Private Rocket Shifts to Island Launch," August 12, 2005, http://www.space.com/missionlaunches/050812_spacex_island.html (accessed October 1, 2006).

192   SpaceX facilities at Kwajalein: Elon Musk, "June 2005 Through September 2005 Update," SpaceX Web site, http://www.spacex.com/updates_archive.php?page=0605–1205 (accessed February 26, 2007).

192   "Tropical hell": Tom Mueller, interview in El Segundo, California, May 2, 2006.

192   Falcon 1's first launch: Elon Musk, "Post-Launch Analysis," Kwajalein Atoll and Rockets, March 25, 2006, http://kwajrockets.blogspot.com/2006/03/post-launch-analysis.html (accessed October 5, 2006); Kimbal Musk, "Someone's Looking Out for That Satellite," Kwajalein Atoll and Rockets, March 25, 2006, http://kwajrockets.blogspot.com/2006/03/someones-looking-out-for-that.html (accessed October 5, 2006).

193   Cause of Falcon 1 crash: "DARPA Approves Return to Flight for Falcon I Launch Vehicle," Defense Advanced Research Projects Agency press release, July 18, 2006, online at http://www.darpa.mil/body/news/2006/return_to_flight.pdf (accessed October 5, 2006); Tom Mueller, interview in El Segundo, California, May 2, 2006, e-mail correspondence, September 20, 2006.

193 Three chances to launch: Elon Musk said he'd probably "throw in the towel" if he had three launch failures in a row, SpaceX press conference, November 18, 2005, from my notes; he clarified that he thought that cancelled launch orders following the failures would probably force him to shut down, telephone interview on January 16, 2007.

193 Pegasus failures: "U.S. Launch Successes and Failures, 1957–1999," Web site of the Aerospace Corporation, http://www.aero.org/publications/crosslink/winter2001/03_table_2.html (accessed October 5, 2006).

Budget Suites of Outer Space

196 Most of the information in this chapter comes from visits I made to Bigelow Aerospace in Las Vegas and from conversations I had there with Robert Bigelow and his engineers on July 29, 2004, and November 2–4, 2004.

197 $18.95 million a month for a stay on a Bigelow space station and the $2.95 million fee for an extra month mentioned later: Robert Bigelow, telephone interview, February 3, 2007.

198 Cost of a typical Las Vegas mega hotel on the Strip: The New York–New York Hotel and Casino cost $485 million, according to a press release from the hotel, "New York–New York Walkthrough," January 18, 2007, at http://www.nynyhotelcasino.com/press_room/press_room_releases_template.aspx?id=117 (accessed February 26, 2007).

200 Bigelow's UFO inspiration: Robert Bigelow, telephone interview, December 30, 2004.

203 NIDS Web site: http://www.nidsci.org/mission.php (accessed September 5, 2006).

203 NIDS as FAA's UFO contact point: Web site of National Institute for Discover Science (NIDS), "NIDS Becomes Only Official Organization to Receive UFO Reports from the Federal Aviation Administration (FAA)," http://www.nidsci.org/news/faa.php (accessed September 5, 2006).

203 Article by Marcia Dunn: Marcia Dunn, "Launch. Inflate. Insert Crew," *Air & Space Smithsonian*, April/May, 1999, p. 20.

204 The habitable volume of the International Space Station was 425 cubic meters (as opposed to 660 cubic meters for two docked BA–330s) in 2006: "The ISS to Date (7/12/06)," Web site of the National Aeronautics and Space Administration, http://spaceflight.nasa.gov/station/isstodate.html (accessed September 5, 2006).

205 Six times faster than an M16 rifle bullet: the muzzle velocity of an M16 bullet is 3,282 feet per second, or a kilometer per second: Web site

of the Navy UDT-SEAL Museum, "M16 Automatic Rifle," http://www.navysealmuseum.com/heritage/armory/m16.php (accessed September 5, 2006).

206 Hard as concrete: Robert Bigelow, *The Space Show*, interview by Dr. David Livingston, August 24, 2006, available online at http://www.thespaceshow.com/detail.asp?q=543 (accessed September 5, 2006).

207 "One of the major breakthroughs in the space program . . ." and other quotes and surrounding TransHab details: Marcia Dunn, "Launch. Inflate. Insert Crew," *Air & Space Smithsonian*, April/May, 1999, p. 20.

207 TransHab: Glen Golightly, "Choosing a Home in Space," Space.com, August 24, 1999, http://www.space.com/news/spacestation/iss_conf_823.html (accessed September 5, 2006); Marcia Dunn, "Launch. Inflate. Insert Crew," *Air & Space Smithsonian*, April/May, 1999, p. 20.

207 Unity node launch date: "Space Station Assembly: Unity Node 1," Web site of the National Aeronautics and Space Administration, http://www.nasa.gov/mission_pages/station/structure/elements/node1.html (accessed September 5, 2006).

208 H.R. 1654: Alex Canizares, "Congress OKs $28.8 Billion NASA Spending Bill," Space.com, http://www.space.com/news/spaceagencies/nasa_spending_bill_001017.html (accessed September 5, 2006).

208 For some good commentary on the space station cost overruns that led Congress to cancel TransHab, see Keith Cowling, "NASA Watch Responds to the Mars Society's Save Transhab Lobbying Campaign—Part 2," June 15, 1999, republished from Cowling's NASA Watch Web site at http://www.hvcn.org/info/a2s2/Zubrin061399.html (accessed September 5, 2006).

208 Text of H.R. 1654, accompanying report, and quote from Congressman Sensenbrenner: Web site of the Library of Congress, "National Aeronautics and Space Administration Authorization Act of 1999, http://thomas.loc.gov/cgi-bin/cpquery/T?&report=hr145&dbname=106& (accessed September 5, 2006).

210 Gibbs's recollections: Franklin ("Gene") Gibbs, telephone interview, November 17, 2004.

210 Vectran in Bigelow's modules and Vectran properties: Leonard David, "Inflatable Space Outposts: Cash Down on High Hopes," Space.com, June 16, 2004, http://www.space.com/businesstechnology/techwed_bigelow_hotels_040714.html (accessed September 5, 2006); "Vectran™ Fiber Introduction," Vectran Web site maintained by Vectran manufacturer Kuraray Co., Ltd. http://www.vectranfiber.com/engineering_introduction.asp (accessed September 5, 2006).

212 "They're taking a very down-to-earth approach . . .": Taber MacCallum, telephone interview, November 18, 2004.

212 "The basic technology is likely to work.": John M. Logsdon, telephone interview, November 22, 2004.

213 Rules of America's Space Prize: this is the full text of the rules available on the Web site of Bigelow Aerospace, http://www.bigelowaerospace. com/multiverse/space_prize.php (accessed September 5, 2006).

217 Launch of *Genesis I*: "Bigelow Aerospace Looks to the Stars," Bigelow Aerospace press release, online at http://www.bigelowaerospace.com/ multiverse/BA_looks_to_stars.php (accessed September 5, 2006).

218 "It felt like becoming a parent": Nell Boyce, "Hotel Tycoon Eyes Affordable Space Vacations," *All Things Considered*, National Public Radio, July 13, 2006, online at http://www.npr.org/templates/story/story. php?storyId=5555718 (accessed September 5, 2006).

218 Bigelow's "Fly Your Stuff" program: Robert Bigelow, telephone interview, June 6, 2006; "Fly Your Stuff," Web site of Bigelow Aerospace, http:// www.bigelowaerospace.com/fly_stuff/ (accessed September 5, 2006).

218 "We are out of the closet. . . .": Robert Bigelow, *The Space Show*, interview by Dr. David Livingston, August 24, 2006, available online at http:// www.thespaceshow.com/detail.asp?q=543.

Spaceport!

220 Proposed spaceports in Singapore and United Arab Emirates: Tariq Malik, "Suborbital Rocketship Fleet to Carry Tourists Spaceward in Style," Space.com, February 22, 2006, http://www.space.com/businesstechnology/060222_techwed_spaceadventures.html (accessed October 24, 2006). Swedish spaceport: Martin Waller, "Space Cadet," *Times* of London, March 22, 2006, Business, p. 63, online at http://business.timesonline.co.uk/ article/0,,8210-2097105,00.html (accessed October 24, 2006). Scottish spaceport: David Lister, "Lossiemouth Reaches for the Stars in Space Tourism," *Times* of London, August 26, 2006, p. 30.

220 Spaceports in New Mexico, Wisconsin, Florida, California, Virginia, Texas: Leonard David, "Spaceports: Building up the Space Travel Industry," Space.com, May 17, 2006, http://www.space.com/businesstechnology/ 060517tech_spaceport.html (accessed October 24, 2006).

220 Sixth largest economy in the world: if California were an independent nation, its gross domestic product (GDP) would rank between France's and Italy's as the world's sixth largest, according to the California Legislative Analyst's Office: "Cal Facts 2004: California's Economy and Budget in

Perspective," http://www.lao.ca.gov/2004/cal_facts/2004_calfacts_econ. htm (accessed October 24, 2006).

221 Mojave fires, other details of Mojave history: John Sweetser, "Condensed History: Mojave Railroad Depots & Hotels," timeline included on Mojave map produced by the Mojave Chamber of Commerce, January 1992.

221 J.W.S. Perry: Mojave Chamber of Commerce pamphlet, "Mojave: California's Gold Crossroads."

222 Uses of borax and details of the twenty-mule-team wagons: Douglas Steeples, "Mules, Mines, and Millions: Frank Smith and Calico Borax," *Montana: The Magazine of Western History*, Spring 2000, online at http://www.findarticles.com/p/articles/mi_qa3951/is_200004/ai_n8896831/pg_1 (accessed October 24, 2006).

222 Mojave gold rush: "History of Mojave Area," included in Mojave map produced by the Mojave Chamber of Commerce.

222 "How many people live in Mojave?" Jeana Yeager and Dick Rutan with Phil Patton, *Voyager*, Knopf, 1987, p. 5.

222 Mojave Airport early history: "Airport History," included in Mojave map produced by the Mojave Chamber of Commerce; "History," Web site of Mojave Airport and Spaceport, http://www.mojaveairport.com/ (accessed October 24, 2006).

223 RAF in Mojave in 1974: Vera Foster Rollo, *Burt Rutan: Reinventing the Airplane*, Maryland Historical Press, 1991, p. 35.

223 National Test Pilot School: National Test Pilot School Web site, http://www.ntps.edu/ (accessed October 24, 2006).

223 Voyager menu: Leonard David, "Spaceport to Rise in California's Mojave Desert," Space.com, May 24, 2004, http://www.space.com/news/mojave_spaceport_040524.html (October 24, 2006).

223 Pegasus: "Pegasus," Orbital Sciences Corporation Web site, http://www.orbital.com/SpaceLaunch/Pegasus/index.html (accessed October 24, 2006); "Orbital Sciences Pegasus," Scaled Composites Web site, http://www.scaled.com/projects/pegasus.html (accessed October 24, 2006).

223 Rotary Rocket: Leonard David, "Rotary CEO Gary Hudson Quits Amid Rocket Delays," Space.com, June 26, 2000, http://www.space.com/businesstechnology/business/roton_rocket_000626.html (accessed October 24, 2006); "Rotary Rocket Roton Atmospheric Test Vehicle (ATV)," Scaled Composites Web site, http://www.scaled.com/projects/roton.html (accessed October 24, 2006); "Roton ATV," Classic Rotors, the Rare and Vintage Rotorcraft Museum Web site, http://www.rotors.org/roton/roton.htm (accessed October 24, 2006); Elizabeth Weil, *They All Laughed at Christopher Columbus: An Incurable Dreamer Builds the First Civilian Space-*

ship, Bantam, 2002; "Rotary Rocket Assets to Be Auctioned," Spaceandtech. com, http://www.spaceandtech.com/digest/sd2001–04/sd2001–04–012. shtml, January 29, 2001 (accessed October 24, 2006).

224 "It's not the edge of the world. . . .": Jeana Yeager and Dick Rutan with Phil Patton, *Voyager*, Knopf, 1987, p. 5.

224 "It's a crummy little desert town . . .": Burt Rutan, address at High-School Senior 4.0 Recognition Dinner, Mojave, California, April 27, 2006, from my notes.

226 Bill Deaver: I met with Deaver at Voyager Restaurant on April 26, 2006.

226 Al and Cathy Hansen: I met with the Hansens at their hangar on April 27, 2006; details of their air and ground vehicles at the Web site of their company, Mojo Jets, http://www.mojojets.com/ (accessed October 25, 2006).

228 Stuart Witt: Stuart Witt, interview in Mojave, April 26, 2006; "Board of Trustees—Stuart Witt," Web site of the Kern Community College District, http://www.kccd.edu/Board%20of%20Trustees/StuartWitt.aspx (accessed October 26, 2006); "Leadership Team: Stuart Witt, Advisor, Airport Operations," Web site of Rocket Racing League, http://www.rocketracingleague.com/bio_stuart-witt.html (accessed October 26, 2007); "Company History," Web site of Computer Technology Associates, http://www. cta.com/index.php?page=history (accessed October 26, 2006); TOPGUN: "NSAWC-Naval Strike and Air Warfare Center," Web site of Naval Air Station, Fallon, Nevada, http://www.fallon.navy.mil/nsawc.asp (accessed October 26, 2006).

230 "Millions for a billionaire": Analysis of California Senate bill number SB 1671 by Jennifer Gress, http://info.sen.ca.gov/pub/05–06/bill/sen/sb_1651–1700/sb_1671_cfa_20060413_160427_sen_comm.html (accessed October 26, 2006).

231 Boyle's report: Alan Boyle, "Spaceport Turnaround," Cosmic Log, MS NBC Web site, http://www.msnbc.com/id/12359455, April 19, 2006, updated noon ET. Boyle revised his report after the spaceport bill was approved by the Transportation and Housing Committee, as Jeff Foust reported in "No 'Millions for a Billionaire' in California," Personal Spaceflight, http:// www.personalspaceflight.info/2006/04/19/no-millions-for-a-billionaire-in-california/, April 19, 2006.

232 Masten's Santa Clara headquarters in Santa Clara and the test site in the hills above: "Test Stand and Infrastructure," January 17, 2006, Web site of Masten Space Systems, http://masten-space.com/blog/index. php?paged=2 (accessed January 28, 2007).

232 Masten's startup date: "Masten Space Systems Is Open for Business," Masten Space Systems press release, August 16, 2004, online at http://masten-space.com/blog/index.php?paged=4 (accessed October 27, 2006).

233 SODASat: "New Service Offers Payloads to Space for $99," Masten Space Systems press release, April 18, 2006, online at http://masten-space.com/blog/?p=75 (accessed October 27, 2006); Masten Media relations manager Michael Mealling, e-mail correspondence, February 6, 2007.

233 Masten Space plan: "About Us," Masten Space Systems Web site, http://masten-space.com/about-vision.html (accessed October 27, 2006).

233 Masten Space Systems history, future plans, and amount of funding: Allison Gatlin, "Masten's Message Simple: Have Rocket, May Travel," *Antelope Valley Press*, September 11, 2006.

233 Masten Space Systems' new headquarters: I visited Masten Space Systems and met with Dave Masten and his crew on July 10, 2006.

234 Lunar Lander Challenge: Michael Belfiore, "One More Giant Step, Please," *Popular Science*, October 2006, p. 37.

234 "How long continuously does your wife cry after seeing Mojave . . .": Burt Rutan, talk at High-School Senior 4.0 Recognition Dinner, Mojave, California, April 27, 2006, from my notes.

238 Masten's demo firings at X PRIZE Cup 2006 and subsequent commercial interest: "January 2007 Update," January 11, 2007, Web site of Masten Space Systems, http://masten-space.com/blog/ (accessed January 28, 2007).

238 New Mexico spaceport press conference and associated quotes: the conference took place on December 14, 2005, in New Mexico, and I teleconferenced in.

239 Futron study: Futron Corporation, "New Mexico Commercial Spaceport Economic Impact Study for State of New Mexico Economic Development Department," December 30, 2005.

239 New Mexico State University study: Arrowhead Center at New Mexico State University, "Business Plan for the Southwest Regional Spaceport," from research conducted in 2005.

242 56-foot-high launch rail: Leonard David, "First Launch from New Mexico Spaceport Nears," Space.com, March 23, 2006, http://www.space.com/news/060323_spaceport_build.html (accessed October 30, 2006).

242 Flight of UP Aerospace's rocket: Science writer Ed Regis was at the launch, and he described the scene in e-mail correspondence, September 27, 2006; "UP Aerospace, Inc. Releases Initial Analysis of Its Rocket Launch from New Mexico's 'Spaceport America,'" UP Aerospace press release, September 29, 2006; KOBTV Eyewitness News 4 report, September 25, 2006,

online at http://www.kobtv.com/viewer.cfm?VID=rocket_fails092506.wmv (accessed October 30, 2006).

The Sky's No Limit

244 Maglev features described: U.S. Department of Transportation Federal Railroad Administration, "Report to Congress: Costs and Benefits of Magnetic Levitation," September 2005, online at http://www.fra.dot.gov/downloads/RRdev/maglev-sep05.pdf (accessed November 19, 2006).

245 The ideas in this chapter about solar power satellites and their receiving stations come from Ralph Nansen, *Sun Power: The Global Solution for the Coming Energy Crisis*, Ocean Press, 2005, and a telephone interview with Nansen on November 6, 2006.

246 Commercial planetary rovers: The ideas presented here are based on plans put forth by David Gump's LunaCorp and described by Rex Ridenoure and Kevin Polk, "Private, Commercial and Student-oriented Low-cost Deep-space Missions: A Global Survey of Activity," paper presented at the 3rd International Academy of Astronautics (IAA) International Conference on Low-Cost Planetary Missions, April 27–May 1, 1998, Pasadena, California, p. 6, online at http://www.smad.com/analysis/IAApaper-finaldoc.pdf (accessed November 20, 2006).

247 Mars mission flight and surface-stay times: Robert Zubrin with Richard Wagner, *The Case for Mars: The Plan to Settle the Red Planet and Why We Must*, The Free Press, 1996, pp 78–80.

248 Commercial flyby mission brokered by Space Adventures: "Space Adventures Offers Private Voyage to the Moon," Space Adventures press release, August 10, 2005.

248 Ninety-nine billion eight hundred million dollars cheaper than *Apollo 2.0:* NASA plans to spend $100 billion on returning to the moon, reported Brian Berger, "NASA to Unveil Plans to Send 4 Astronauts to Moon in 2018," Space.com, September 14, 2005, http://space.com/news/050914_nasa_cev_update.html (accessed November 20, 2006).

248 Lunar landscape 62 miles below: "What Will I Experience," Space Adventures Deep Space Expeditions Web site, http://www.deepspaceexpeditions.com/section3-DSE.html (accessed November 20, 2006).

249 Not much more required for commercial moon mission: Eric Anderson, Space Adventures CEO, described the larger view ports for me in a telephone interview on August 11, 2005. The camera mounts are my addition.

250 "Based on our research and investigation . . .": F.W. Smith, "Practical

Applications of Hypersonic Flight: Possibilities for Air Express," Battelle Institute, 1986, quoted by Robert Zubrin, *Entering Space: Creating a Spacefaring Civilization*, Jeremy P. Tarcher/Putnam, 1999, p. 50.

250 A shut-down assembly line could cost a manufacturer $200,000 an hour: J.C. Martin and G.W. Law, "Suborbital Reusable Launch Vehicles and Applicable Markets," prepared by the Aerospace Corporation for the U.S. Department of Commerce Office of Space Commercialization, October 2002, online at www.nesdis.noaa.gov/space/library/reports/2002–10-suborbital-LowRes.pdf (accessed November 20, 2006). This figure is given for the real-life downtime cost of the computer chip maker Intel. This report also gives an overview of another potential benefit of affordable routine access to space: the zero-gravity development and manufacture of biotechnology and inorganic materials. Protein crystals, metal alloys, components like ball bearings, and other substances can all be produced with greater purity and precision in the absence of gravity-induced convection currents and the need to place an object on a surface while it is formed. The commercial viability of this application of the space environment probably won't be fully understood, however, until it is actually available at a reasonable cost to industry, so I haven't addressed it in this chapter.

250 Commercial Space Transportation Study (1994): online at http://www.hq.nasa.gov/webaccess/CommSpaceTrans/ (accessed November 20, 2006). This study gives the four-hour figure for a human heart to survive outside the body and gives the circuit board example as an item worth shipping by spaceship.

250 Ansari on the overview effect: Anousheh Ansari, "Watching the World Go By," Anousheh Ansari Space Blog, September 26, 2006, http://spaceblog.xprize.org/2006/09/26/watching-the-world-go-by/ (accessed November 21, 2006).

251 Ansari on Oprah: The *Oprah Winfrey Show*, ABC television network, October 14, 2006, online at http://video.google.com/videoplay?docid=−8201115994630204234 (accessed November 21, 2006).

252 New energy source criteria: This is a variation on five criteria outlined by Ralph Nansen in *Sun Power: The Global Solution for the Coming Energy Crisis*, Ocean Press, 1995, p. 76. Nansen's criteria are that the new energy source be 1. Low cost; 2. Nondepletable; 3. Environmentally clean; 4. Available to everyone; 5. In a usable form.

252 Ralph Nansen: The details of Nansen's career and his work on the space shuttle definition and solar power satellites come from Ralph Nansen, telephone interview; November 6, 2006, *The Fourth Era*, un-

published book manuscript, 2006, Chapter Six, "What If?"; and *Sun Power.*

252 Space shuttle early design and production contracts: Cliff Lethbridge, "History of the Space Shuttle Program," spaceline.org, http://www.spaceline.org/rocketsum/shuttle-program.html (accessed November 24, 2006).

253 F–1 rocket engine thrust rating: David West Reynolds, *Apollo: The Epic Journey to the Moon*, Tehabi/Harcourt, 2002, p. 86.

254 Build the shuttle with no budget increases: Dennis R. Jenkins, *Space Shuttle: The History of Developing the National Space Transportation System*, second edition, self-published, 1992–1997, p. 107; Cliff Lethbridge, "History of the Space Shuttle Program."

254 Frank Moss's involvement with the shuttle's solid rocket boosters: Mark Wade, "SRB," Encyclopedia Astronautica, http://www.astronautix.com/engines/srb.htm (accessed November 25, 2006).

255 NASA centers against the fly-back booster, Nansen's reaction to NASA dropping the fly-back booster: Ralph Nansen, telephone interview, November 6, 2006.

255 Nansen reassigned to program-cost analysis: Nansen, *Sun Power*, pp. 11–13.

255 "Will the person leaving SEATTLE . . .": Ralph Nansen, *The Fourth Era*, Chapter Six, "What If?"; Greg Lange, "Billboard appears on April 16, 1971, near Sea-Tac, reading: Will the Last Person Leaving SEATTLE—Turn Out the Lights," June 8, 1999, HistoryLink.org: The Online Encyclopedia of Washington State History, http://www.historylink.org/essays/output.cfm?file_id=1287 (accessed November 26, 2006).

255 "Big Onion" dialogue: Nansen, *Sun Power*, p. 12; telephone interview, November 6, 2006.

256 Peter Glaser's 1968 paper: Peter E. Glaser, "Power from the Sun: Its Future," *Science*, vol. 162, Number 3856, November 22, 1968, pp. 857–861. This paper was adapted from one Glaser delivered at the Intersociety Energy Conversion Engineering Conference in Boulder, Colorado on August 13, 1968.

257 Wireless power generation first proposed by Nicola Tesla: Peter E. Glaser, "Japan—The 21st Century's Global Energy Supplier?", *Journal of Practical Applications in Space*, 1993, online at spacefuture.com, http://www.spacefuture.com/archive/japan_the_21st_centurys_energy_supplier.shtml (accessed November 27, 2006).

257 William Brown's demonstration of wireless power transmission: Richard M. Dickinson, "Bill Brown's Distinguished Career," Web site of Institute of Electrical and Electronics Engineers (IEEE) Microwave Theory and

Techniques Society, http://www.mtt.org/awards/WCB's%20distinguished %20career.htm (accessed November 27, 2006); W. C. Brown, abstract for "Experimental Airborne Microwave Supported Platform," report number RADCTR–65–188, December 1965, online at the Web site of the Defense Technical Information Center (DTIC)'s Scientific and Technical Information Network (STINET), http://stinet.dtic.mil/ (accessed November 27, 2006).

257 Boeing's solar power satellite study contract and Nansen's involvement: Nansen, *Sun Power*, pp. 13–18.

257 Boeing's solar power satellite technical details: Nansen, *Sun Power*, pp. 20, 187, 204–207.

257 Experimental power transmitter and receiver in the Mojave Desert: William N. Agosto, "Astronaut Plugs Solar Power Satellites at 1976 International Microwave Symposium," *L5 News: A Newsletter from the L–5 Society*, number 11, July 1976, online at www.nss.org/settlement/L5news/ L5news/L5news7607.pdf (accessed November 27, 2006).

258 "I have a standing offer ...": J.P. Smith, "Industries Seek Billions through Solar Satellite," *Washington Post*, April 30, 1978, Sunday, Final Edition, p. A10.

258 Solar power rectenna would absorb 99 percent of radio energy: Nansen, *Sun Power*, p. 207.

258 Greenhouses suggested by John J. Olson: Nansen, *Sun Power*, p. 208.

258 Space solar power abandoned in 1980, Nansen's recollection: Nansen, *Sun Power*, pp. 25–28, 212.

258 Hundreds of launches required for solar power satellite: Nansen, *Sun Power*, p. 22.

259 1993 suborbital power transmission test: Nansen, *Sun Power*, p. 133; "ISY-METS Rocket Experiment," Web site of the Research Institute for Sustainable Humanosphere at Kyoto University, Japan, http://www.kurasc.kyoto-u.ac.jp/plasma-group/sps/mets-e.html (accessed November 28, 2006).

259 Powersats invisible from Earth except at night, maintenance requirements, longevity: Nansen, *Sun Power*, pp. 189–190, 194–196.

259 Hydrogen production: Jeff Wise, "The Truth About Hydrogen," *Popular Mechanics*, November 2006, online at http://www.popularmechanics. com/technology/industry/4199381.html (accessed November 28, 2006).

260 "Space solar is ridiculous": Elon Musk, telephone interview, January 16, 2007.

260 "SSP offers a truly sustainable, global-scale and emission-free energy resource": Martin I. Hoffert quoted by W. Wayt Gibbs in "Plan B for Energy," *Scientific American*, September, 2006, p. 108; Hoffert's other opinions: Marty Hoffert, telephone interview, January 15, 2007.

261 Air and Space Museum's nine million visitors a year: National Air and Space Museum Press Kit, "Overview," Web site of the Smithsonian National Air and Space Museum, http://www.nasm.si.edu/events/pressroom/presskits/museumkit/overview_nasm.cfm (accessed November 28, 2006).

261 *sic itur ad astra*: "Thus one goes to the stars," Virgil, *Aeneid*, book nine, line 641.

Index